Energy for Rural and Island Communities

Energy for Rural and Island Communities

Proceedings of the Conference, held in Inverness, Scotland
22-24 September 1980

Edited by
JOHN TWIDELL
University of Strathclyde
Glasgow, Scotland

PERGAMON PRESS
OXFORD · NEW YORK · TORONTO · SYDNEY · PARIS · FRANKFURT

U.K.	Pergamon Press Ltd., Headington Hill Hall, Oxford OX3 0BW, England
U.S.A.	Pergamon Press Inc., Maxwell House, Fairview Park, Elmsford, New York 10523, U.S.A.
CANADA	Pergamon Press Canada Ltd., Suite 104, 150 Consumers Rd., Willowdale, Ontario M2J 1P9, Canada
AUSTRALIA	Pergamon Press (Aust.) Pty. Ltd., P.O. Box 544, Potts Point, N.S.W. 2011, Australia
FRANCE	Pergamon Press SARL, 24 rue des Ecoles, 75240 Paris, Cedex 05, France
FEDERAL REPUBLIC OF GERMANY	Pergamon Press GmbH, 6242 Kronberg-Taunus, Hammerweg 6, Federal Republic of Germany

First edition 1981

British Library Cataloguing in Publication Data

Energy for rural and island communities.
1. Power resources - Congresses
I. Twidell, John
333.79'11'091732 HD9502.A2 80-41995

ISBN 0-08-027290-8

In order to make this volume available as economically and as rapidly as possible the authors' typescripts have been reproduced in their original forms. This method unfortunately has its typographical limitations but it is hoped that they in no way distract the reader.

Printed and bound in Great Britain by
William Clowes (Beccles) Limited, Beccles and London

Preface

'Need' and 'opportunity' are the key words to describe energy supplies for island and rural communities in the 1980s. The need is to cater for the continuation and expansion of supplies at a time when costs from centralised sources are rising above inflation. The opportunity is to be the first to benefit from the modern development of renewable energy sources, which must first take place on a small scale at sites favouring these sources. Likewise where costs are rising steeply, there is the strong desire to improve the efficient use of energy with the opportunity to utilize modern developments in building and energy conservation.

The conference was planned from the Energy Studies Unit and Department of Applied Physics at Strathclyde University because the Highlands and Islands of Scotland were believed to show both the need and the opportunity. Our original ideas were for a small working seminar to consider a few local problems, but interest increased rapidly to warrant activity on a larger scale. It became apparent that whilst a single island or rural community might be treated as an oddity of no significance within a nation, there were so many thousands of similar communities on a world scale that the need of one becomes the need of many.

The distinctive nature of the conference was to value small scale development and to stress realistic application. Self sufficiency and self reliance became frequent themes. We wished to create a down-to-earth approach, and were pleased at the mix of participants from industry, business, regional administration, research institutes, universities and the general public. A small exhibition of wind, hydro and wood-burning devices, and energy efficiency applications helped to stress practical application.

By the conclusion of the conference it was apparent we had made a start along the road of rural and island energy development. The complexity of the opportunities for new energy sources was seen as support for self sufficiency and local development, which certainly included many ideas for business and increased employment. Both technical and economic difficulties were fully realised, and there was unanimous agreement about the poverty of resources being allocated to solve these difficulties. Nevertheless the opportunities are so great, that there was a general mood of optimism.

The mix of subjects covered was not fully comprehensive, yet papers ranged from the biological to engineering, from financing a small business to environmental impact. Energy supplies could certainly not be considered in isolation from each other, and an optimum mix of supplies and resources could lead to robust, stable and sustainable communities in contrast to the uncertainties of recent times. Interest was also shown by those favouring large scale development of renewable resources, since small communities provided the opportunity to begin realistic development. This is obvious for instance regarding the introduction of wind generators in the U.K., that the Scottish islands provide excellent sites both for wind potential and local need. It is probable that a similar opportunity exists for the development of relatively small scale wave power systems.

The conference is to be repeated in 1981 on an extended scale, and will again be organised from the Energy Studies Unit of Strathclyde University. Eden Court Theatre makes an excellent site for the conference with the facilities for exhibiting equipment within and outside the building. Our policy is to include meals taken in the Theatre dining room as part of the conference programme, and to arrange accommodation in the hotels and boarding houses along the River Ness, within walking distance of the Theatre. A policy of moderate and interrelated energy supply should be advocated by a conference displaying similar aims in itself, and the environment of Inverness allows this without hardship and indeed with much benefit. Our thanks are due to the management of Eden Court Theatre, and to all the participants and speakers who so enthusiastically supported the conference.

John Twidell
Energy Studies Unit and Department of Applied Physics,
University of Strathclyde,
Glasgow.

Contents

TOPIC A
Strategy and Action

Energy for Rural and Island Communities—An Introduction

John Twidell

Department of Applied Physics and Energy Studies Unit,
University of Strathclyde, UK

ABSTRACT

Rural and island communities form the major proportion of World population, yet receive little attention for energy development. Only the harnessing of the dispersed low-intensity renewable forms of energy, in small scale local application, will provide energy for these communities. Development is complex, but must be based on ecological principles. Modern science and technology has much to offer in harnessing renewable supplies, but the associated affects of local development and employment must be considered. Examples are given from the conference programme to reinforce the appreciation of these principles.

KEYWORDS

Renewable; energy; rural; technology; planning; development.

INTRODUCTION

Energy Needs

The rural population of the world forms 60% of the total 4,400 million population. The fraction is decreasing as people in the Third World are drawn to cities for employment, social services, economic markets and energy supplies. This urban drift is seen as a major difficulty in development for universal prosperity, and all countries have agencies to reinforce rural life. In Scotland all this has been seen before, so that now the Highlands and Islands area with 17% of Britain's land area has but 0.6% of the population. The number of people economically maintained in the area is moreover much less than the whole due to retirement and unemployment. We may compare this with 200 years ago when the same area maintained six times the proportion of Britain's population. The reasons for the changes are complex and often beyond control, yet energy supply clearly plays some part. At this conference we shall concentrate on energy supply as a poorly understood and even less developed aspect of rural and island life.

We must adopt a down-to-earth practical approach to the problem. There are too many academic conferences and publications on national large-scale energy strategy that have little relevance for rural and island communities. We are told there are no energy difficulties now, and the main need is to consider insurance for the future rather than tangible developments for immediate use. This is however poor advice, and such statements are too easily made by national authorities working in urban administrations out of touch with rural and island life. In Britain, such policies stem mainly from South East England.

A Lesson from Fiji

Recently I had the privilege of working in the Fiji Islands of the South Pacific. Within the programme of rural electrification, islanders were given diesel generators for cooperative village grid networks on the condition they maintained the supply themselves. Their difficulty however was how to pay for the fuel and services even if the capital payments were donated. Oil price was like a rising meniscus that was steadily submerging development and threatening established livelihood. At the time it appeared to me that this meniscus was rising through the less developed world, and that it would be many years before it engulfed the developed countries. I was wrong. On returning to the U.K. I find the identical problem is being faced by certain Scottish islands. The only difference is that the islands in Scotland are being treated as minor oddities with little national impact, whereas the Fiji islands are credited with national importance representing a problem being faced by 60% of the World's population. The common link is the extraordinary power of small islands to act as microcosms for the analysis and solution of problems that have immense complexity on a national scale.

There is a further aspect of these island electric energy supplies that is of major importance. The difficulties appear to be economic, but are fundamentally technical. Present energy efficiencies range between 10% and 30%. This is far too low, since it is quite feasible for efficiencies to rise close to and beyond 100% by controlling all aspects of the energy and using heat pumps. Moreover fuel costs can become zero if renewable sources are used. In practice therefore energy development and application is 'device limited' and not 'cost limited'.

Energy for Whom?

We are considering energy <u>for</u> rural and island communities. We are not considering the extraction or release of energy <u>from</u> rural sites to be transmitted to urban complexes. This is an important difference and we must be clear about the distinction. It is this aspect that distinguishes our conference from many others, and with which we must hope to produce real and immediate benefits for rural people. I believe the distinction will lead us to realise that <u>small scale renewable sources of energy</u> offer by far the best opportunities for improvements, especially when local industry and employment opportunities are included in the equation. It is people that matter, not things. It is pointless to send energy to a rural area if the population has evaporated. Energy developments demand large resource allocations, but all too frequently the money for capital and maintenance appears as benefit outside the rural area of concern and does not reach the local community. We know it is the local education and employment of young people that give vigour to a community, and unless a full range of job skills can be exercised the community will decline. Thus I am sure we shall be keeping an eagle eye on energy developments that lead to increasing local employment and manufacture in our rural and island communities. This is more likely to happen by harnessing the dispersed renewable forms of energy and using these at low intensity in the vicinity, rather than bringing in supplies from centralised sources or concentrating the renewable supplies for export to urban complexes. Several of the papers at our conference will be describing commercial and planning activities stemming from energy developments, and I hope such subjects will feature strongly in discussion.

ENERGY REQUIREMENTS

Rural and island areas of the world are now critically dependent on oil for commercial energy supplies. The extent of this dependence is still not fully appreciated at the local level, and it is urgent that we quantify these energy flows. The way we use energy now relates to the life we wish to preserve, and we must reinforce this life before considering new developments. We have one conference paper describing a complete energy survey of an island community in Orkney, which will remind us how we use energy. Perhaps this will convince every local administration to conduct energy surveys as preparation for energy developments.

Energy requirement may be likened to a multilayer cake that becomes sweeter and more decorative to the top, but relies on hidden foundations below which must not crack. For most of the world the main foundation is firewood for cooking and space heat. This is usually a non-commercial energy flow of extreme importance. In Scotland this relates to peat and the yet unquantified supplies of waste wood from forestry, and falls within the study of biomass production and the use of waste products. We shall be having several papers within this general area, at both a strategic and local industrial level.

Heat

In cold and temperate regions a basic requirement for energy is space heat, and in all agricultural and manufacturing industry there is need for process heat. On the islands of Northern Scotland domestic heating is continuous through the year and forms about 50% of total energy demand. Architectural design must allow for this, and we have several papers relating to building construction. These show that in Scotland it is surprisingly important to design for passive solar heat gain and essential to regulate ventilation. When considering alternative supplies of energy, the need for space and process heat usually dominates over the need for controlled electricity. Thus it becomes sensible to consider direct energy conversion to heat from wind and hydro sources.

Transport

Haulage and transport are dominant energy requirements in rural and island areas. Again it is necessary to quantify the needs and consider local effects. Any fuel system based on large-scale oil refining or synthetic petroleum production from coal will prove expensive for these areas. Located at the outer circumference of any centralised energy supply, the communities will bear the greatest distribution costs, will appear of minor market importance, and will not benefit from the cash flow and employment opportunities of centralised production. For instance in Northern Scotland there is stark contrast between the difficulties of rural and island energy supplies and the energy bounty of the off-shore installations feeding centralised refineries.

There is but one strategy to lessen the problems of transport to rural and island areas - the strategy of local self sufficiency for goods and of modern telecommunications for information. The former must certainly include food, and the latter can today include two-way video and facsimile telecommunication. Within the communities, alternative forms of transport such as electric and hydrogen powered vehicles are exceptionally well suited for local use. The requirements for short predictable range, back to base operation, low speed, robust construction, and local maintenance expertise are all met in agricultural and island use. In addition there is ample opportunity for local wind or hydro sources to charge the energy store provided as part of the vehicle. We may further note that these vehicles may be 'fueled' at night and so provide a load to balance daytime electrical use.

Electricity

Heat and transport dominate as energy requirements in the rural areas of temperate and cold latitudes. If we include the provision of frozen food from refrigeration as an aspect of heat, then we can safely say that heat and transport are dominant requirements throughout the world. What is the importance of electricity therefore as an energy supply? The answer is that electricity is of prime importance as a supply of high quality power for high quality living. We all expect to have as a public service, electricity for lighting, refrigeration, washing machines, power tools and entertainment. Nevertheless there are important qualifications to make.

(a) Electrical services form an icing on the energy cake - they are attractive, useful and acceptable only so long as the foundation supplies of heat and transport are maintained at prices that can be afforded.

(b) Grid electricity supply from centralised intensive sources is by no means the only way to supply electricity.

(c) Electrical supply can be integrated with other forms of energy supply and storage to improve overall economics, as in combined heat and power installations and wind power integration into established networks.

(d) Very often electricity supply can ride 'piggyback' on dominant supplies such as space and process heat. The cost of the electricity is then a marginal cost on another installation, and so the unit costs of electricity can be much reduced. For example lighting from batteries becomes attractive in association with wind machines built for heat production.

(e) 'Electricity' must not be used as a synonym for 'energy'. Thus quite misleading policies can result from using the term 'nuclear energy' for 'nuclear grid electricity', and island communities mislead themselves if they consider all energy needs are satisfied by a submarine cable connection to a mainland grid.

We shall be having several papers that touch on these themes, but it is to be regretted that opportunities for integrated energy supplies are so handicapped by the present policy of separated services. This not only limits supply, but limits initiative and incentive.

ENERGY COMPLEXITIES

Energy is not a subject for study by itself. We have to ask what energy does for us and how it does it. We find that a correct flow of energy leads to a satisfactory and sustainable quality of life through mechanisms of great complexity.

How are we to know what is correct however? Do we have any examples or principles to guide us? Fortunately there are both examples and principles.

Natural systems are maintained by the flow of dispersed low-intensity solar energy that continuously cycles organic material. The systems are complex, have long term stability, and are sustainable. Man can interact with these systems and adjust them for his advantage, but only within certain limits. We are more wise to cooperate with nature therefore than to fight against it. Our greatest experience is in agriculture where sound ecological principles are respected and used. It would be folly indeed for rural and island communities to forget the importance of agriculture, and it is important for technological development to respect these ecological principles and adapt to them. Technology must learn to harmonise and not be discordant with nature.

Examples

Consider the example of a combined heat and power generator fueled by methane from a biological waste digestor. The machine may be classed as a 'final stage parasitic decomposer' in a trophic system driven by solar energy. Classification of this sort describes the full worth of the device, and is far better than merely acknowledging the machine in money saving terms. An advertising agency might disagree however.

There are other complexities of which the relationship of employment to energy supply is most important. It is urgent that local authorities consider this question with all seriousness since the answers are not obvious. Consider some examples from the Highlands and Islands of Scotland.

(a) Colonsay now has a local cooperative electricity supply from small diesel sets not connected to the national grid. There is part employment for a skilled crofter as a half time engineer, for a merchant who handles the fuel, and for the ferry. The system also leads to local initiative and challenge for social cooperation. However for various understandable reasons costs per unit of electricity produced are high and the system must change as we shall be hearing in a paper at this conference. There are but two strategies for improvement: either (i) the system must head for a complex self-sufficient operation involving such aspects as combined heat and power, automatic load control for optimum operation, the inclusion of transport and the incorporation of wind and other renewable sources of power, or (ii) electricity should be supplied by submarine cable from the national grid and no thought taken for integrated energy planning. The best answer to the choice is not obvious, but I find it disappointing that the decision with regard to electricity was not openly debated in these terms.

(b) The Highlands and Islands are famous for a biological product - whisky. For fundamental reasons of processing and marketing the distilleries will always be based in rural and island environments, but in a modern distillery the number of employees has been reduced to a minimum. Even so this employment is vital to the localities and intricately bound to the biological process. If however the distillery is seen as part of an ecological system using biological materials efficiently and avoiding any waste of energy, then much development is possible. There is the obvious possibility of linking horticulture to the distilling activities. One important result is the resultant increase in local employment at the skilled and semi-skilled level. We shall be having an example of such a distillery described to us.

(c) The widespread use of hydro power in Scotland seems an obvious advantage to us now, but was not so obvious in the 1940's when Mr. Tom Johnston, the Secretary of State during the Second World War, had the vision to establish the North of Scotland Hydro Electric Board. We will appreciate the importance of hydro power when I remind you of the generating costs of public power in Scotland last year, as taken from the official reports. The figures are pence per kilowatt hour for 1979/80

	North of Scotland	South of Scotland
Hydro	0.70	0.26
[Steam (coal, oil), Gas turbine, Interchange]	2.00	1.72
Diesel	3.66	-
Nuclear	-	1.75

Whilst some of the hydro plants are not the most beautiful buildings, there is general public satisfaction with the environmental aspects of the systems, and there is confidence in the associated employment for hundreds of years to come.

Not all hydro power is arranged by the Public Boards. Private generation is well established, with great potential for many thousands of small scale installations. We shall be hearing details of one such installation that has become associated with developments for local industry, and other papers will detail various device possibilities. There are also clear indications here of small firms manufacturing the generating equipment in the hydro areas - an example that should surely be followed by manufacturers of small scale wind power equipment.

Energy Agrarians

Recently a name has been suggested for people who recommend that commercial energy supplies should follow natural processes - 'energy agrarians'. This is a reasonable name, but it fails to describe the engineering challenge of most of the devices and systems. It is quite misleading to suggest that any ideas in alternative energy are so easy that homesteaders can knock up the required gadgetry using back yard scrap. The task of extracting reliable supplies of energy from a wild and fluctuating environment is a daunting task. The supply moreover must fit the loading and dynamic characteristics of a modern society wishing to live independably of their environment. This is a new challenge for modern engineers requiring new science and new ideas. It is not a return to old fashioned engineering since so much more is known today about materials and control systems. Modern microelectronics for instance allows a completely new range of systems to be designed to increase the efficient use of energy. This could not have been contemplated in the 1930's.

Modern Science and Technology

Thus the most modern scientific techniques are needed to consider and develop energy for rural and island communities. This will not happen unless resources are allocated in the same way as any other development of national importance, and unless the very special challenges of the rural environment are appreciated. For instance there have been enormous national and international resources put into the development of nuclear electricity. It is entirely reasonable to expect corresponding resource allocations to renewable energy and energy efficiency. However allocation of money is not sufficient by itself; the institutions and industries that the money encourages must be established in the correct places. Consider wind energy. This is a complex subject ranging from meteorological measurement, through structural engineering, to the incorporation of the supplies into community use. There is a strong case for the UK, which is one of the world's windiest countries, to have a National Wind Energy Research and Development Authority. This should be placed in a region where the wind is available, the technology welcomed and the supplies needed. Mainland Orkney is such a place, but South East England is not. Yet I doubt whether such a development will occur in a rural or island environment unless the local population press the case. It is all too easy for the pressure of national administrative convenience to place such institutions near urban cities, notably London. One can be quite sure that energy will not be available for rural and island communities unless the appropriate planning authorities press for development in their own localities. Neither the centralised UK Department of Energy in London nor the Energy Technology Support Unit, placed within Atomic Energy Authority at Harwell, have the stimulus or vision to tackle what they see as a minor problem. This requirement for local planning authorities to be drawn into the energy debate adds yet another complexity to an already scientifically complex subject. Yet there is no alternative, and for this reason we plan to conclude our conference with presentations and discussion on alternative energy and local development.

CONCLUSION

It is easy to dream dreams, but action following dreams requires more care. My dream is for a world of self sufficient rural and island communities more prosperous than their urban counterparts. Communities where traditional values are respected, and the young remain to enrich and encourage new development. Can the dream be realised? If so, one requirement is undoubtedly the need to harness local renewable sources of energy and use these efficiently and carefully. The energy is there in abundance within the complexities of the natural environment, but neither science nor local industry has developed sufficiently to produce reliable supplies.

Our task at this conference is not unlike that of a ship's captain. The energy ship has been steering on a course set some time ago, and is now heading for the rocks. We have to make two types of decision, firstly to set a new long range direction perhaps completely different from that previously, and secondly to manoeuvre round individual rocks in the short range so as to come about into the new direction. It is folly to avoid immediate rocks (such as diesel oil surcharge) if there is no long term strategy (such as developing wind energy), and it is equally folly to look only to the long range direction (such as the UK wave energy programme) without avoiding errors near at hand (such as lack of ceiling insulation). The ship can sail through, but the best of science and technology must be used by a crew sensitive to the needs of the passengers.

Energy for Remote Communities—The Strategy

Malcolm Slesser

Director, Energy Studies Unit, University of Strathclyde, 100 Montrose Street, Glasgow G4 0LZ, UK

ABSTRACT

Energy is an indispensable element of running a modern economy. The more centralisation, whether of Government or industry, the more the periphery is rendered dependant on the decisions, whims or needs of the heavily populated centre. The basic strategy for the peripheral communities is to adopt life-styles, methods of work, energy provision and energy-use systems that render them increasingly independant of the centre. Renewable energy has just such a potential. Rural communities can capitalise on their greater land availability per capita. However it will need positive government help, both with capital for systems and modified legislation to allow a greater degree of adaptation to local circumstances.

KEYWORDS: Energy; strategy; independent; government.

The poet Donne said "No man is an island". We might paraphrase that to say that no physical island is an economic island. That is every community, isolated or otherwise, draws some succour and inputs from outwith its system. The problem arises when the external system imposes on the internal. The fact is, of course, that the evolution of a highly centralised economy in the UK with its rigorous and vigorously enforced rules of safety, with an all-embracing social security system and with unionisation of labour permeating to the furthest corners of the kingdom, what are norms, feasible and even desirable in the urban/industrial zones can become burdensome to remote communities and reduce their viability. Part of the strategy must be to allow the people in remote communities more freedom to adapt to their circumstances rather than forcing them to conform to urban norms. After all, it was these genetically evolved powers of adaptation and will to survive that led to those personal qualities we all so much admire, and which must inevitably be sapped undermined by legislation adapted for the urban hinterland. I make these points because I believe we cannot, either technologically or economically, solve the problems of remote communities without also relieving them of some of the burden of over-weaning centralisation of thought and action.

Let me now move to the wider system, the world beyond the UK's shores. Where lies the world's energy future? You will find many opinions expressed, and little guidance as to which are credible. Moreover, the economics profession is in confusion over the significance of energy prices on other costs, and this confusion makes it impossible to interpret a forecast, even if one believed it. Rather than give you a distillation of the views of others, I give you my own analysis, based on many years experience in this field, and a very quantitative approach to the analysis of the data.

There is no impending energy crisis in the sense of running out of energy. The world's energy stocks are simply enormous. However they are getting harder and harder to get at, and as time passes each additional barrel of oil, each additional kilowatt of electricity will require behind it so much more capital, and just so much more primary energy input for each unit of useful energy output. You might look upon it as an energy inflation within the energy system dictated by the laws of thermodynamics. The situation at the present time is worse than it need have been because of the failure of those in power (or, shall we say, their advisors) to invest in the new energy systems before the existing one's like readily available oil from the middle east became expensive. This failure to appreciate what was happening arose out of the fact that economic theory has no paradigm for dealing with energy as one of the worlds two non-renewable resources (the other is time). It was therefore left to those who saw energy resources through the second law of thermodynamics to offer a word of warning, but it went unheeded to a great extent, and still does today.

We are in for a period when, in order to secure our energy supplies quite enormous investments in the energy sector are going to be necessary. Already these run, globally speaking, at over 25% of total world capitalisation, and by the end of the century, will be running at about 45%. What we spend on securing our energy supplies will, inevitably, reduce what we can spend on hospitals, schools, universities or industrial expansion. It will prove a very hard time for governments, for people have been conditioned over the last few years to expectations which cannot be fulfilled any longer for reasons largely beyond the immediate control of government.

The present price of energy is absurd and had steps been taken using other energy sources to prevent oil becoming a market leader, it could have been avoided. In 1970, when there were almost as many energy options open to us as now (since then solar energy has made big strides, nuclear has not), investments on alternative systems were deferred because they were ten, fifty or a hundred percent more expensive. Now we are finding oil 1500% more expensive and little control over its price. It is worth noting that the UK government chose an expensive energy policy.

In my view, on the long haul, oil will rise on average about 30% a year till such a time as OPEC oil is no longer a market leader. It will drag with it an inflation rate almost as high, mitigated by our skills at learning to use energy better. For the third world and for island and remote communities this is a sombre forecast. The direct energy costs of supporting island communities are substantial, higher than those of supporting an urban community. If these communities are obliged by the nature of government and the economic system to import the means of their survival from the economic heartland, then they are doomed. Island and remote communities will be refuges for the rich only.

But the very fact of rising fossil and fissile energy prices, and the trend of energy conversion technology towards larger and larger units, offers to open a price window for decentralised energy systems. Before embarking on this theme, let me first ventilate two schools of thought that exists here, which in turn draw their conclusions from differing views of how energy prices affect producing costs. The first view is that since energy represents but a mere 6% or so of manufacturing and producing costs, the impact of a big rise in energy prices, though undesirable, are not grave. For example a doubling simply makes the 6%

into : $\frac{2 \times 6}{106} = 11.3\%$

This view ignores any gearing effect as energy prices move through the economy. The other view, which takes this into account, argues that in the absence of a fall in labour's buying power and without a move towards better use of energy, a doubling in energy prices will double costs, but with a time lag of some five years. This view seems more consistent with observed facts, but it is too early to be sure. I shall not argue the case further, but simply point out that it has important repercussions on investment decision making. If the first view is correct, then capital intensive energy technologies such as nuclear or solar have only to bide their time, and they will become viable. Yet though energy has increased 1500% in price since 1973, it is hard to find any which have become more viable. If the second view is correct, (or more nearly correct, because labour has lost buying power and there is a small move towards better energy use), alternative energy systems become viable only when their technology has improved. Nevertheless, there is a time window after an energy price increase, when existing capital equipment is cheaper than it will be in the future - hence there are opportunities for investment which would not exist in a stable price environment.

What energy does.

We cannot work out our strategy until we have a clear idea of what it is energy does for our lifestyle and standard of living. Man utilises two quite distinct qualities of energy. The familiar one is for heat - warming homes and offices. The second is for doing work. It may not be generally appreciated that the reason we enjoy a materially, more affluent lifestyle than our forefathers is because we have many slaves to work for us. In the UK, each citizen has about fifty slaves at his behest. They are in the form of machines which do work, and need energy to operate them. Just imagine how many people would be needed to row a MacBraynes steamer from Uig to Stornoway, or a sleeping car from Edinburgh to Inverness - it would be late even more often! Associated with every use, there is a capital structure and an energy source. It may be readily shown that economy of energy goes hand in hand with greater investment, provided the investment takes into account natural limitations. An example of a poor, indeed stupid misguided investment has been the development of evermore powerful fishing trawlers, so that the energy use per unit of fish caught has soared to the point of economic disaster. This is the direct result of unfettered competition for a finite resource. A sensible approach is to restrict the catch or the catches and adjust the method to the right combination of capital and energy use. Figure 1 shows this event in terms of the Scottish herring fleet before and after the last war.

The scope for energy conservation in buildings is enormous. It would be safe to say that practically every house built since the brick and metal era commenced uses three or four times as much energy as an efficiently designed (but more expensive) home. It may be readily shown that, within a very short time, investment in conservation pays off. Such conservation in island communities reduces imports. Government subsidy should be for conservation rather than freight subsidies to imported fuel for there is then less to freight.

The next step is to use the natural advantages of the remote communities. These are, first and foremost, space - a lot of land per person. Solar energy capture is land intensive. Secondly, and this especially is true of island communities, there is much wind. We need a creative approach to rendering an island steadily more energy independent. Inevitably, this means looking at the community as a whole, not just the technology. There is no point in having locally produced energy, if minor maintenance cannot be handled locally. This is either achieved by abandoning demarcation in work, or by maintaining or building up the communities to a viable size, with a range of skills.

All over the world there is a move towards the use of renewable energy sources for self reliant development. As members of a developed country, our own remote communities are entitled to be helped along that path, freeing them from the economic shackle of centralist sources. Already so many opportunities have been missed. Let us take the recent example of Skye, where the solution chosen was to lead pylons from the mainland to the island. Not only was an eyesore created, but a golden opportunity in a windy isle to develop and sustain a indigenous energy source, with fruitful activity for local people. It would, of course, have been more expensive in terms of immediate capital. But with rising energy prices, that original investment would soon pay its way, and provide a design guide for others. Coll is another example where an under-water cable will bring in mains electricity whereas the same large capital investment could develop a local renewable source of energy.

Finally, renewables are not the only answer. There are devices, such as the small combined heat and power units (e.g. Totem by Fiat) which make better use of fossil energy. There are more sensible ways of doing many things . There is no future along a path which suggests we do without energy - that is merely to put the clock back. Rather, going forward with new technology for dispersed energy production, hand in hand with encouragement to develop and initiate a self reliant lifestyle on islands and remote communities, unhampered by the mores of the urban centres is the way to protect then develop remote communities.

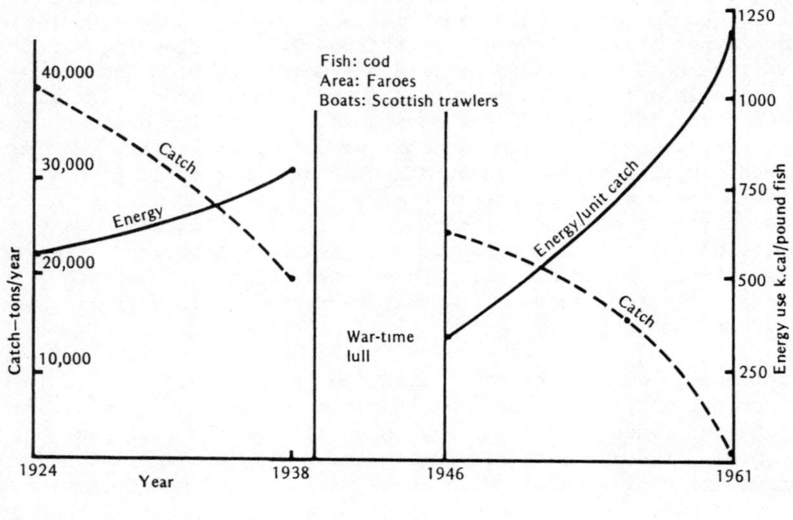

Fig. 1. Energy consumption in deep-sea fishing.

Sea Wave Energy for Rural and Island Communities

Thomas L. Shaw

Sir Robert McAlpine & Sons Ltd., London, UK

1. INTRODUCTION

Electricity accounts for about one-third of the energy we consume. Sea waves may more readily be converted to electricity than to any other generally useful form of energy (hydrogen may in time prove an exception to this). The value of wave energy depends upon the contribution it may make to the daily and annual pattern of electricity demand. This Paper assesses the form of this contribution, especially to communities isolated from major power networks.

The so-called 'oil crisis' of 1973/74 fortunately prompted the U.K. and other Governments to reappraise all energy sources at their disposal, especially those occurring within its national territory. A number of countries looked at wave energy, especially Japan, Norway, South Africa, U.S.A. and U.K.

Waves had been little studied prior to these investigations, yet for the U.K. and for many other countries this is one of the larger renewable sources. It has so far attracted a substantial share of the funds earmarked for the renewables, and it could be said that these have been sufficient to stimulate adequate ideas and basic information to judge its real potential.

If this is so, we may reasonably expect soon to be at the point at which at least a first comprehensive evaluation of the source may be made. Should that prove sufficiently favourable against criteria that must recognise the fact that it is still early in the process of development and demonstration of a new technology, the next step would include the preparation of schemes on a pilot scale that will, coincidentally, be appropriate to the demands for energy by remote communities.

It is therefore timely to report on progress with studies of the potential of this source to the present audience.

2. THE NATURE OF WAVE ENERGY

A primary objective for the electricity supply industry is that it shall supply all our needs for electricity upon demand with a high degree of reliability. Many industries have to accept this specification. However, no other industry also has the tasks of creating and supplying their commodity at the rate and time required by the consumer, because unlike other products, electricity cannot readily be stored in bulk.

The real value of wave energy depends upon the efficiency with which it may be harnessed, converted and transmitted to the consumer, i.e. the correspondence that exists between its availability and the electrical demand that has to be met. The degree of this equality, through each day and from day to day through typical years, will therefore relate to the timescales by which both the wave energy resource and the pattern of consumer needs vary and in particular how they differ from each other. The nature of the latter is much better known than the former, in fact the availability of wave energy seems hardly to have been evaluated at all. To exploit any energy source demands that equipment is designed to operate at its best efficiency at one specified energy level, or 'rating'. The choice of this level is critical to the total operating economics of the power station. It is most easily selected for thermal stations such as coal and nuclear where it corresponds to the level at which one would, if possible, run the station continuously over periods of months and even years, plant performance permitting. However, all the renewable sources are to varying degrees, intermittent. Geothermal is probably the most persistent, while the tides may be the least. The timescale of wind energy is also short; wave energy is rather better, for the reasons discussed below.

'Persistence' is not, of itself, an essential criterion. There must also be a use for the energy when it can be produced, otherwise it will have no value. Real value depends not only on the pattern of consumer demand but also on the cost of meeting that demand in all possible ways. The wave energy contributing to the cheapest form of supply is that which has value.

The cost of not being able to use a renewable (or any) source at any time because it is not available must also be recognised in its evaluation. Other plant must then be called up, which means that further capital must also be committed to cover this situation. If the renewable source is very persistent yet still not totally reliable, the standby plant would seldom be used. However, it must nevertheless be constructed and held at readiness according to the probability that the renewable source will fail to function for any reason. The unit cost of the electricity it produces will then be particularly high though its annual fuel bill will be small.

All power stations, including fossil and nuclear, have to be backed-up by reserve plant, the % being fairly well known for all thermal generators. This is an expensive element in all cases and is a continual target for economy within the limitations of maintaining an adequately firm supply. The preferred way of minimising it depends upon the type and size of each power station in the system. Reliability of the principal source is an obvious asset here, that is it must be both technically and operationally (fuel supply) dependable. A well proven generating station fed by a guaranteed fuel source and able to operate efficiently, economically, on demand and at any level of output is therefore an ideal that, in practice, is most closely approached by hydro-electric plant. This does not mean that other sources must be less satisfactory than hydro, though most if not all, are.

What also matters is how well all sources may be combined together to give an optimum combination of reliability and economy. Without a specification of the proportion and type of all other plant that could operate in any necessarily integrated electrical supply system, hence the back-up plant this will need, it is very difficult to do more than identify the likely generating and cost criteria for each energy source according to the operating experiences available for each.

To the characteristic of 'persistence' must therefore be added those of 'reliability' and 'suitability', i.e. we must know the extent to which wave energy is able to contribute to electrical networks fed in varying degrees by this and by other sources and the measure of firmness or guarantee that must then be attached to each contribution.

It is inappropriate to attempt to foresee here the broader circumstances that will determine whether or not a wave energy station will prove economic in future supply situations. The main reason for this is that techniques for harnessing the source have still to be proven at full scale, hence it is not yet possible to know with certainty how much energy will be produced from a sea state of specified roughness. In turn, this means that the optimum rated output of a wave energy scheme cannot be selected, hence its design and cost cannot be drawn up. Nevertheless, there is ample evidence from those energy and offshore engineering markets with which wave energy development has much in common to expect that the economic status of this developing technology may already be judged with a precision that fairly reflects both its embryo nature and its prospects.

From the resource view-point, there are adequate data to indicate the persistence of waves at most offshore locations at which devices might be sited, i.e. the probability that the density of incoming energy will exceed any specified threshold for any number of hours (or days) at any time of the year. These statistics are not yet generally available though their preparation is a straightforward task. It is also an essential exercise for it shows the merits and disadvantages of combining a wave energy station into any electrical network.

Wave energy has its origin in the wind, i.e. in solar energy. The process of wave generation by wind action is one of progressive energy transfer, manifested as growth of wave height and length (period). It takes time for substantial wave energy to accumulate, though once this has occurred we have in wave action one of the most efficient stores in terms of its ability to retain the amassed potential, as well as a resource that naturally moves, again with high efficiency, from the region in which it is accumulated to that close to coastlines where it may best be harnessed.

The sea therefore acts as an areal integrator and reservoir of wind energy. It has the ability to transport energy from storms occurring in different areas to a common location. Larger sea areas experience more storms, each of which will extend over a greater area and last for longer before dying away or passing onto an adjacent land area. The downwind coastal area will, in these cases, experience greater and more persistent wave action, i.e. the extent of the adjacent sea area and the probability of storms of any duration, strength and frequency occurring over it will determine the magnitude of the available energy resource.

Scotland is particularly well placed to benefit from this process, on both a small and a large scale.

The delay between the occurrence of storm wind action and the arrival of the consequential wave energy at the coastal location is, for an ocean, typically of the order of a few days. The intensity and time of arrival of this energy can now be forecasted adequately accurately from satellite observations for this to be accorded a substantial measure of firmness. Indeed, this is one of wave energy's strongest advantages. The forewarning available is in keeping with the timescale needed to adjust the output of thermal stations so that the combined outputs of the two are integrated together efficiently. Also, the relatively long duration for which the density of wave energy exceeds any threshold and the slow rate at which that density varies make this a very suitable source to supply to any electrical network.

A further advantage of wave energy is that its mean monthly density varies seasonally in approximate accordance with the variation in our demand for electricity. Winter levels are substantially higher than those in summer, though in this respect one cannot but notice that the extreme energy levels are also many times higher than the longer term averages. In economic terms these only have a negative value because while the structures must be designed to withstand the severe forces involved, the energy density will then be much in excess of economic plant ratings, hence will not appear in the energy budget of the station.

The energy density of sea waves is proportional to the product of the square of wave height and the wave period. The proportionality factor depends upon the extent to which the wave motion is regular and directionally mixed. In general terms, the higher are the waves the more mixed is the sea, but the energy potential is then greater and hence is more likely to exceed the economic rating of the power conversion system. Present information suggests that waves of height 3 metres and period 10 seconds will give an input close to the economic threshold of wave energy conversion systems. The energy of higher and longer waves should therefore be captured with less efficiency (which is technically almost inevitable) in order not to overload the transmission system. The capacity of this should be equivalent to about 50kW per metre of wave front captured, i.e. 50MW per km. A power station of 1000MW capacity would therefore cover 20km of sea length.

For the conditions off the west coast of Scotland such a station would operate at or close to full capacity for a large proportion of the winter months. Its persistence, reliability and suitability would be well suited to electricity supply.

The situation during the summer months would be less favourable to maintaining a firm supply but this need not be a serious limitation. Its economic consequences will depend on the characteristics of those other stations that may then be made available to the system.

3. COSTS

It is too early to make firm estimates of the capital, maintenance and running costs of wave energy plant, though detailed figures should be produced within the next 12 months. However, at that stage they will not be based on experiences with large scale experiments in the sea. That essential step must follow quickly if the first cost estimates, based on all relevant offshore operating experience, supports the case for prototype studies. If that step is warranted, it could involve construction of units rated at up to a total of about 10MW, and its justification would be essentially experimental. However, it would be logical for it to supply a real load close to its chosen location, and this could be in Scotland. A successful scheme could not only be the forerunner of a much larger installation supplying the national network but it could also be reproduced at similarly small scale elsewhere to supply other localised and isolated demands.

Reference was made in Section 2 to the cost penalty to be carried if extensive reserve plant is needed to back-up intermittent primary sources. This offsets the benefits of harnessing 'free' energy sources. Although it is not a penalty unique to the renewables, the size of it could then be substantial if the renewable sources are expected to meet a large proportion of total demand.

The possible ways of minimising this penalty, for example by including energy storage facilities in the system or by reducing the guaranteed level of supply to customers, are of great practical importance though secondary here to the principal point that, for small communities, the renewable sources would in many cases displace a proportion of a particularly expensive supply if that comes from oil. The disadvantage of additional capital for reserve plant is then correspondingly reduced, though the benefit of the renewable source is still maximised when its reliability is high and in sympathy with the needs of the consumer.

Now that oil-based generating costs from large new stations are likely to exceed 7p/unit and from small stations in remote areas are said to be over 10p/unit, it is very probable that wave energy will be able to compete on economic terms. The detailed economic case remains to be completed for the basic reasons given earlier but the evidence to date suggests that supply costs lower than those for oil will be achieved.

A particular problem for wave energy systems is that of survival in storm conditions and the cost penalty that has to be paid to secure this fundamentally important feature of this type of station. Large scale prototype demonstration schemes would have this as a principal objection for study and rationalisation.

4. CONCLUSIONS

The steady development of wave energy technologies is approaching a stage where prototype demonstration schemes become an essential component to advancing understanding on many topics affecting the design, cost and performance of optimum systems. Plant with a rated capacity of a few MWs, perhaps up to 10MW will be needed for this purpose and it would be logical to locate this where it could supply an isolated community or industrial demand.

Present evidence suggests that the unit cost of electricity produced by a small wave energy station would compete favourably with oil-fired plant supplying an isolated demand, and the characteristics of its operation will be favourable for this purpose. A principal function of a demonstration scheme must be to show how best the extreme forces occurring in severe sea states may safely be overcome with minimum capital cost disadvantage and acceptable security.

ABSTRACT

Many remote communities are well placed to benefit from the energy of sea waves. This Paper considers the merits and disadvantages of this source in terms of the surety of supply it offers and the cost of extracting it relative to the use of diesel generators. It is noted that the high cost of oil and advances in wave power technology now makes it timely to give serious consideration to the renewable source.

Utilisation of Waste Heat in a Malt Distillery

Harry Cockburn

Glengarioch Distillery Co. Ltd.,
Oldmeldrum, Aberdeenshire AB5 0ES, UK
A wholly owned subsidiary of Stanley P. Morrison Ltd.,
13 Royal Crescent, Glasgow G3 7SU, UK

ABSTRACT

The scheme developed at Glengarioch Distillery was to re-use low and high grade heat produced in condensing of distillate and in direct firing of pot stills.

Low grade heat at 71°C, in the form of hot water from still condensers, is employed to supply the heat requirements of a 0.7 hectare horticultural project. High grade heat at 400°C, in the form of combustion products from the stillhouse flue, is employed to produce process steam in a waste heat boiler and for kiln drying of malt.

The horticultural project started in 1976 with a 0.2 hectare greenhouse producing a long season tomato crop for sale in Aberdeenshire. Cucumbers, aubergines, peppers, lettuce, tulips, bedding and potted plants have been grown. Experimentation in various crops has proved that tomatoes must be the main crop.

Our experience is that the minimum size of such a project must be 0.6 hectares. Starting with the 1981 season, we will have this facility. The development of this area produces sufficient income to justify a return on investment after paying for horticultural expertise.

This development has resulted in the employment of three full time and five part time staff.

KEYWORDS

Local market; choice of site; choice of crops; high grade heat; low grade heat; management.

DISTILLERY PRODUCTION

Glengarioch Distillery, situated in the village of Oldmeldrum, 27 km North West of Aberdeen, was established in 1788. It was acquired by our company in 1970 and together with Bowmore Distillery on the island of Islay, it forms the production facility for a company with interests in Whisky broking and blending. The company also owns the Machrie Hotel and Golf Course on Islay and could be classed as a small private company in the Scotch Whisky trade. Both distilleries are termed "traditional" in that they have on-site facilities for the production of malted barley but they vary in that the stills at Bowmore are heated indirectly by steam coils while those at Glengarioch are heated directly by the combustion of commercial butane. In virtually all other respects, they are comparable.

The distillery process can be broken into five distinct areas: *malting*, *mashing*, *fermentation*, *distillation* and *warehousing*.

Malting is the modification of starch and the production of natural enzymes carried out by firstly steeping the barley in water and thereafter allowing the barley to grow under controlled conditions. The malting process is completed by kilning the malt using peat.

Mashing is the conversion of the modified starch in the malt to sugar using the enzymes. The malt is first ground and then mixed with hot water to eventually produce a sugar solution called worts.

Fermentation is the conversion of those sugars contained in the worts to alcohol by the addition of yeast. CO_2 is also produced as a bi-product of the fermentation.

Distillation is the separation of alcohol from the fermented liquor. Two distillations are required to produce the traditional spirit called Scotch Whisky.

Warehousing completes the production cycle. By law, Scotch Whisky must be matured in oak for a minimum of three years before it can be called Scotch Whisky. In practice, three years is a minimum.

LOW GRADE HEAT - THE HORTICULTURAL PROJECT

Why, then, did we choose Glengarioch Distillery and not Bowmore Distillery for the horticultural project? There are many reasons, but the most salient are:

Availability of Land and Buildings

Glengarioch Distillery has an area of approximately 16.4 hectares of land immediately adjacent to the stillhouse. The remnants of a biological tower previously used to treat distillery effluent now house the office, crop grading and packing facility, chemicals and machinery.

Proximity of Markets

As previously stated, Aberdeen is 27 km away and has a wholesale market supplying a local population of some 0.5 million people. Islay has a population of some 4,000 people and has the normal transport problems associated with an island.

Less Exposed Site

The site at Glengarioch has a South facing aspect and although exposed to the prevailing wind from the South West, it is less exposed than an island location. Apart from the obvious possibilities of wind damage, the heat loss from the greenhouse rises proportionately to the windspeed.

Need for Re-cycling of Cooling Water

Aberdeenshire does not have a plentiful supply of fresh water and to reduce the usage of water, a cooling tower was installed for re-circulation. The increase in production at Glengarioch made it necessary for us to consider purchasing a second cooling tower. Our first greenhouse was an alternative to a second cooling tower.

On consideration of the foregoing points, the decision was made for us. We started the horticultural project at Glengarioch Distillery in 1976 when a 0.2 hectare multi-span plastic house was erected. In 1979, a further 0.1 hectare was erected but this time a glass sheeted house was chosen. This second house is used primarily as a propagation house, servicing the existing plastic house along with a further 0.4 hectare glasshouse which has just been erected.

Heating System

The water for the complete project is taken from the still condensers directly to a 54,000 litre storage tank so as to provide a reservoir of hot water to cater for the interruption of supply due to the batch process nature of distillation. From here, it is pumped round a ring main from which it is supplied to each of the individual house heating systems as required. The returned water from each house is then fed back to the cooling tower and thence to the distillery.

The heating pipes in the first two greenhouses constructed are of a conventional type having pairs of heating pipes spread to suit a tomato crop. The heating systems are designed to maintain "blue print" temperatures using condenser water at 71°C. The temperature drop is around 22°C.

The heating system in the new 0.4 hectare house is a combination of a grid system on the floor to supply 40 per cent of the heat load and a pattern of twenty fan-assisted radiators to supply the other 60 per cent. The designed temperature drop is from 71°C to 49°C in the ground pipes and from 49°C to 32°C in the fan system. The temperatures in all the houses are controlled automatically by a sensor which operates a valve to vary the flow of water through the heating system (see flow diagram - page 6, Fig. 1.).

Nutrient and Irrigation System

At ground level in each house, the nutrients are supplied to each plant by a 1 mm bore pipe serviced from a ring main. This feeding is controlled automatically to suit plant demand depending on rate of evaporation. Each house has at high level fresh water sprays which are used to damp down the plants, increase humidity and assist in the pollination of the plants. This system will also be used in the future for enriching the atmosphere with CO_2, a bi-product of our fermentation. This aids the photosynthesis and results in heavier crop yields. So far, we have been reluctant to use the gas as we are investigating the possibilities of the carry-over of bacterial infection and other chemical compounds which could have a detrimental effect on the plant life.

Ventilation

The ventilation of the plastic house is effected by fans ranged along the gable end. These fans are initiated automatically in sequence depending on the rate of ventilation required. This rate is dependent on solar gain and humidity. The signal is transmitted by a sensor in an aspirated screen so as not to be affected by direct sunlight. Ventilation of the glasshouses is also automatic and is in the form of conventional ridge ventilation.

Growing Systems

It cannot be assumed that by supplying heat only, one creates the ideal horticultural environment. Equally important is the value of the light factor. We found it necessary to supplement the hours of light available during the early stages of the plant development. The seeds for long season crops are sown in November so that the early spring market can be met. The additional light is supplied by high pressure sodium lamps which not only extend hours but also improve the quality of the natural light available.

The plants at Glengarioch are sown in peat modules in preference to cultivation of the ground so that the season can be extended to an eleven month growing year. If the soil were to be cultivated, this would necessitate sterilising of the soil to prevent the build-up of various disease mechanisms. These modules are relatively easily removed and replaced and, in fact, they have a second-hand value. The third choice, NFT, was considered and once we have gained more confidence, we would not discard it as a future development.

Financial Implications

When one enters into a new area, one has to anticipate that research and development may cost money. This was true of our tomato project. However, as the years have passed, our experience has been growing. 1979 saw us making a profit on tomatoes but a loss on tulips. 1980 sees us concentrating on tomatoes with a second greenhouse coming into operation; all indications are that a modest profit will be made this year. With the completion of our project in 1981 (with 0.7 hectares in full production), we expect to crop over 130 tonnes of tomatoes and our financial projections for this scheme allow the following forecasts:

TABLE 1 Projections for Horticultural Project

	1981	1982	1983
	£	£	£
Sales	96,670	111,000	128,000
Running Costs	51,665	57,975	65,055
Interest Charges	28,489	28,489	28,489
Depreciation	11,757	11,757	11,757
	91,911	98,221	105,301
Net Profit	£4,659	£12,779	£22,699
Investment at 1 January 1981	£117,570 (net of Grants)		

The projected cash flow from the horticultural project would indicate a positive cash flow by the end of 1983 of £75,418 after interest charges. This would indicate that the payback period in crude terms would be approximately the fifth year.

No allowance in these cash projections has been given for further escalation in oil prices which, we would imagine, would see a rise in horticultural prices. The return in five years is therefore rather pessimistic and in normal circumstances, we would look upon this as a marginal investment.

HIGH GRADE HEAT - WASTE HEAT BOILER

As previously stated, there exists another source of waste heat in the distillery. This is in the form of combustion gases from the still flues at 400°C. These gases are fed to a purpose built waste heat boiler where steam in generated at 16 bar. In the absence of flue gases, the boiler can be overfired with butane to maintain a stand-by facility.

The exit gases from the waste heat boiler at 150°C are fed to the malt kiln obviating the need to burn butane.

The steam from the waste heat boiler is used in the following way: to preheat charges to the pot stills; to heat water for the mashing process; to steam sterilise fermenters; to supplement heat from condenser water to greenhouses.

For comparison, the heat content from the waste heat boiler accruing to the distillery amounts to 300 tonnes equivalent of butane per annum, or £45,000. To heat 0.7 hectares of glasshouse amounts to 230 tonnes equivalent of butane per annum, or £35,000.

SUMMARY

As distillers, we have enjoyed our involvement in horticulture and we have learnt by our mistakes. Inexperience led us to expect too much too quickly. We solved this problem by employing expertise at management level and in our opinion, there is no other way.

The return on investment as mentioned is five years, even with grants included. This might not be attractive to an industrialist. A development of this nature would have to be well supported by some Government agency. Horticulturalists traditionally accept much lower returns bearing in mind that in Scotland they are small family businesses.

Co-operative ventures could be the answer in rural areas with the owner of the heat source contributing free heat as his investment and the horticulturalist giving his expertise as his contribution. The project would have to be jointly financed. Even then, there are pitfalls.

Future technology may render today's good ideas obsolete. One of the major waste heat sources in island areas is the electricity generating station (diesel). 1 MW installed capacity would supply 0.3 hectares of glasshouse. However, should electricity supply companies, having encouraged joint investment in horticulture, develop a cheaper means of supply, the advantages of waste heat would end. Similarly, distillers presently investing money in fish farming and horticulture are still investigating alternative methods of utilising waste heat. Joint ventures would therefore require long term commitment.

We as a company made a commitment to horticulture and with the geographical siting of our complex in Aberdeen, we feel that we have made the right decision. There is a responsibility on us all to investigate the utilisation of waste heat sources and it is my belief that individual solutions can be found for individual situations.

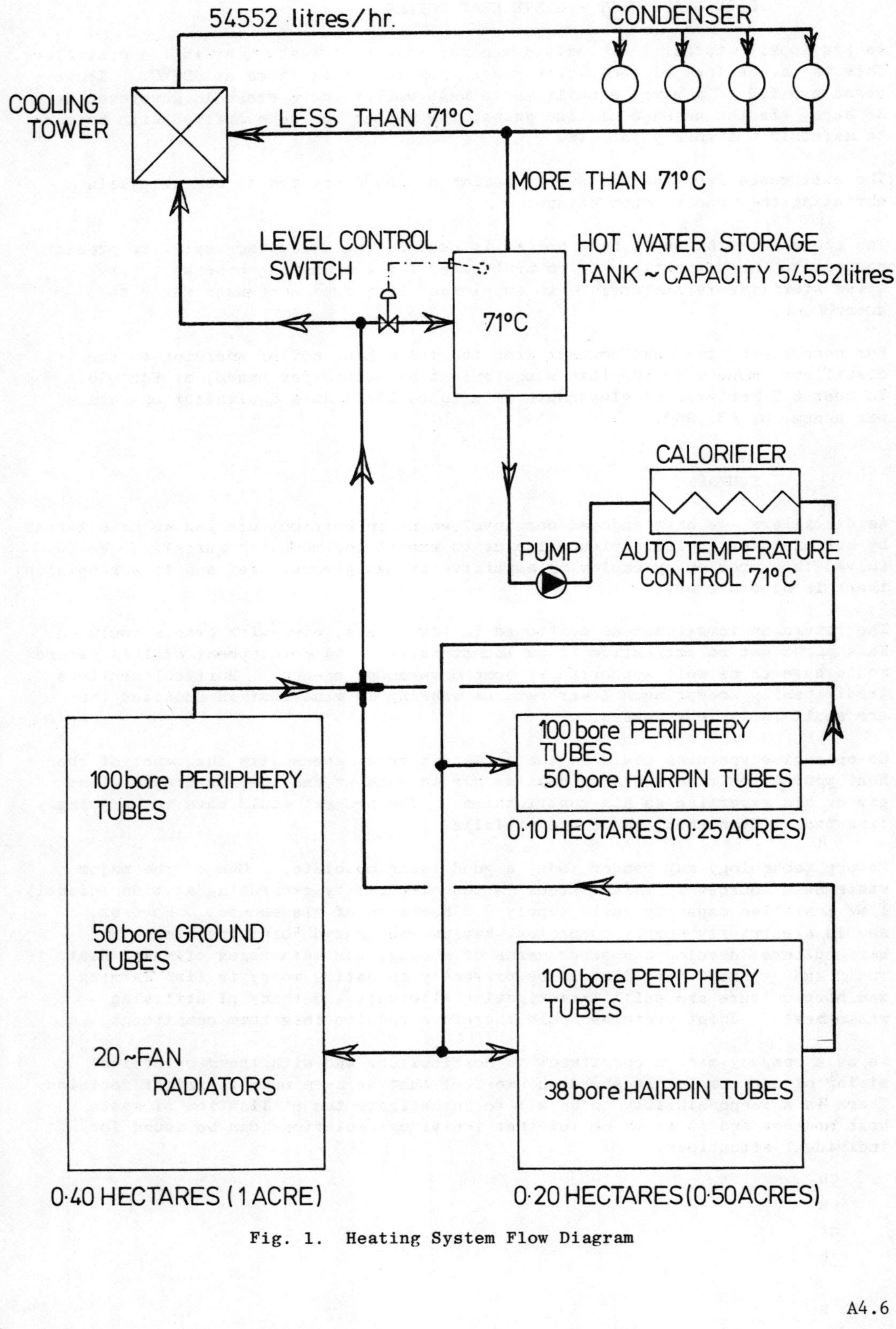

Fig. 1. Heating System Flow Diagram

Cogeneration of Heat and Power— A Market Opportunity

Thomas H. Smale

Fiat Energy (UK), Fiat Auto (UK) Limited,
Great West Road, Brentford, Middlesex, UK

INTRODUCTION

As a result of research and development work and extensive analyses of market attitudes and market research data FIAT has entered the business of energy conservation - the more effective and efficient use of energy. This new venture is an extension of its existing energy business for which it is better known e.g. the building of nuclear power stations, hydro electric schemes, etc.

This paper will explain the introduction of the Total Energy Module - Totem, a cogenerator of heat and power, into the UK market and why we believe it is relevant to the theme of this conference.

TECHNICAL DETAILS

Power Unit and Generator

The Totem consists of a FIAT 127 engine, modified to run on gas fuel and at a constant speed, which is used as the power unit to drive an electric generator. The mechanical arrangement follows the standard pattern used for power generators with the additional feature that all the heat generated by the IC engine and the electric generator is captured by a set of heat exchangers which make the heat available in the form of hot water at 85°C flowing at 2,000 litres per hour. The flow rate can be variable from 1,000-6,000 litres per hour.

FLOW RATE	MAX. RETURN TEMP.	Δ T	OUTPUT TEMP.
1,000 litres/hour	55°C	30°C	85°C
2,000 litres/hour	68.5°C	16.5°C	85°C
3,000 litres/hour	70°C	11°C	81°C
4,000 litres/hour	70°C	8.3°C	78°C
5,000 litres/hour	70°C	6.6°C	76°C
6,000 litres/hour	70°C	5.5°C	75°C

The FIAT module used for the combined production of electric power and heat has been designed around the performance characteristics of the 903 cc 127A engine coupled to an asynchronous type electric generator and a set of heat exchangers for recovering the heat produced.

Operating Features

The operating features of the Totem are:-

Fuel	-	Natural Gas Manufactured Gas Liquid Petroleum Gas (LPG) Biogas Alcohol
Heating Performance	-	Maximum operating temperature for heating purposes: T Max. = 85°C
Power Performance	-	Connected to a 3-phase mains, 415 volts (± 10%), 50 Hz
Noise Level	-	65dBA max.
Emissions	-	Comparable to a standard gas boiler
Safety	-	Fully sensorised and controlled against any thermal or power system failures
Reliability	-	Comparable to a standard boiler

The IC engine operates with a wide open throttle at 3,000 rpm and requires no regulator as it is constantly loaded by the asynchronous generator.

The engine operation is controlled by sensors; they are the:-

- Oil pressure gauge on the input circuit
- Oil level sensor on the auxiliary oil reservoir tap
- Water temperature sensors on the engine head and on the gas/water heat exchanger
- Overspeed sensor

The engine can operate on a continuous or on an on-off basis for a total period of more than 10,000 hours before a major inspection is required.

The scheduled maintenance inspections are at 1,000-1,500 running hour intervals.

The standard version of the Totem (version 5) must be connected in parallel to the national grid. When switched on the power generator (acting as a driving motor) will reach a steady speed on 0.3 to 0.8 seconds therefore the electric power drawn from the mains is limited both in amplitude and time. With the star delta starter system the starting current is 70 amps.

Control Features

The Totem has a control system which fulfils specific functions.

The Totem's self-diagnostic system ensures:-

- a correct thermo-mechanical operation of all its components
- unit shutdown before any failure can damage it
- monitoring the need for maintenance operations

The Totem's control system ensures:-

- a correct connection of the Totem to the power mains
- unit shutdown in the case of power mains failure
- compliance of the unit's operation to thermostat or remote control signals and manual signals

The Totem system is a modular one. One control panel can control up to eight Totems. As each additional Totem is added to the installation so a printed circuit card is inserted into the control panel thereby connecting that Totem into the whole system.

In the second quarter of 1981 the Totem version 6 will be available in the UK. It will be version 6 standard and version 6 independent. The main features of the two new models are:-

Standard

- the width of the Totem has been reduced to 90 cms
- the electronic control system has been redesigned

Independent

In addition to the above

- a starter motor, ignition system and battery built-in

The independent Totem should be considered as a cogenerator of heat and power in exactly the same way as the standard version, connected to the grid as before with the additional features being:-

- the electronic control system monitors the grid, should the grid fail the control system will switch the Totem to an "independent of the grid role", the Totem will continue to produce heat and power. When the grid power comes back on the Totem will switch back to a grid connected status.

- the independent Totem can operate without being connected to the grid.

This independent version of the Totem will therefore have the advantage of being able to operate in a multi-role mode or in those areas where there is no grid supply.

TOTEM APPLICATIONS

As a result of our extensive market research programme we are satisfied that Totem has applications across the range of industry, the commercial sector, public sector, particular applications in the domestic sector and in the agricultural sector.

In the market research work eight general types of Totem applications were used for assessing the market potential.

- Used in conjunction with an anaerobic digester
- Using fuel produced as a by-product in a production process
- Used as an integral part of a production process
- Using Totem with the mains gas supply in competition to other more conventional sources of heat and electricity
- Used in applications with no electrical supply
- Used for peak lopping of electrical demand
- Used as a standby facility

It is interesting to note that as at the end of June 1980 when a major market research report was produced the field research, covering over 250 respondents, strongly indicated that most organizations had yet to size up to a proper view of energy conservation. Even energy managers of major corporations had not been able to develop fully the concept of pay now (for efficient energy conversion equipment) and thereafter reap future benefits. Thus the researchers found themselves researching the market for a product which is to be sold as a solution to a problem which many potential customers had not yet realised existed. However, user appreciation is strengthening as energy prices outrun inflation.

Totem in Conjunction with an Anaerobic Digester

As energy costs spiral upwards the fermentation of organic wastes to produce methane will become increasingly attractive, economically. Totem as a gas engine burning methane is highly suited for use in conjunction with a digester.

Waste materials in general represent an excellent untapped source of stored energy. It is estimated that approximately 90 million tonnes (dry weight) of organic residues are produced every year in the UK - on farms, in forests, in industry and in the home. These residues contain energy equivalent of 60 million tonnes of coal (15% of the UK energy demand) and are potential sources of usable fuel.

The most abundant organic wastes are:-

- sewage sludge
- farm wastes
- domestic refuse

the latter being the largest single source.

Each of the above sectors are regarded as prime markets for the Total Energy Module - TOTEM.

We also regard the Distilleries sector as a potential market there being 129 whisky distilleries in the UK.

Totem with By-product Fuel

The use of Total Energy Modules with fuels produced as a by-product in an industrial process has special relevance in applications using methane from the coal mining process; using methane generated in particular sections of chemical plants.

Totem as an integral Part of Production Process

In industry Totem will be of special interest to companies which have demand for both hot water and electricity in their manufacturing processes.

Totem in Competition with Conventional Units

For Totem to be competitive with conventional sources of heat and electricity the increased capital cost of Totem must be justified by the primary energy input savings.

As an indication we give some typical payback figures based on 1980 industrial prices i.e. electricity from the mains at 3 pence per kWh; mains gas at 27 pence per therm; Totem maintenance costs of 6.25 pence per running hour (all of which are conservative figures). For a domestic customer Totem is more attractive, provided the load factors are correct since domestic electricity is 4 pence per kWh while natural gas is 21 pence per therm. (Prices in remote or island communities being higher.) Figures quoted are average and depend on gas usage levels, gas contract, etc.
The definition of load factor is the running time as a proportion of the year.

TOTEM COMPARED TO BOILER + MAINS ELECTRICITY (Examples from a comprehensive table)

Number of Totems	Boiler	Totem Running Costs per hour (fuel cost & Maint.) £/hour	Total Boiler (Gas) & mains Elect. Costs £/hour	Capital Cost of Boiler & PV of Maint. over 5 years £	Capital Cost of Totem & Control Unit /£
1	King-fisher 125	0.60	0.90	600	3750
2	Avon 1/8	1.21	1.79	1130	7000
3	Diplomat 2/18	1.81	2.70	1050	10250

Approximate excess capital pay-back periods can be calculated at various load factors. For example:-

One Totem at 30% Load Factor

Excess capital expenditure of Totem over boiler	3750 - 600 = £3150
Totem running costs/hour saving	0.90 - 0.60 = £0.3
Hours to payback excess capital expenditure	3150 ÷ 0.30 = 10500
Payback period at 30% load factor	10500 ÷ (0.30 x 8760) = 4 years

A payback period of four years for equipment of this type is sometimes considered too long for industrial concerns. However, the payback period would decrease as the load factor increases. The table below shows the effect of load factor on payback.

Typical Payback Periods - Totem versus conventional Boiler plus Mains Electricity

Load Factor	Typical Payback Periods in Years (Number of Totems)							
	1	2	3	4	5	6	7	8
10%	12.0	11.5	12.0	11.3	9.4	10.7	10.5	12.3
20%	6.0	5.8	6.0	5.7	4.7	5.3	5.2	6.1
30%	4.0	3.9	4.0	3.8	3.1	3.6	3.5	4.1
40%	3.0	2.9	3.0	2.8	2.4	2.7	2.6	3.1
50%	2.4	2.3	2.4	2.3	2.3	2.2	2.0	2.4
60%	2.0	1.9	2.0	1.9	1.6	1.8	1.7	2.0
70%	1.7	1.7	1.7	1.6	1.3	1.5	1.5	1.8
80%	1.5	1.4	1.5	1.4	1.2	1.3	1.3	1.5
90%	1.3	1.3	1.3	1.3	1.0	1.2	1.2	1.4
100%	1.2	1.2	1.2	1.1	0.9	1.1	1.0	1.2

Totem providing Main Electricity Supply

The following sectors where there is no mains electrical supply were investigated:

- remote houses, farms, hotels
- leisure centres in remote locations, e.g. trekking centres, camping sites

Although this sector of the market is obviously unsuitable for the Totem Mk I version (i.e. the present standard version) it is suitable for the Totem Mk II version (i.e. the independent version) available in 1981.

In this market sector the potential user of Totem has a requirement for a unit providing heat and/or electricity which could be met by one of the following:

- Boiler - gas, oil, solid fuel, etc.
- Generator set - diesel, petrol, gas, etc.
- Combination of a boiler and generator set
- Cogeneration units similar in concept to Totem

BOILER/GENERATOR SET COMBINATION COSTS (Examples from a comprehensive table)

Equivalent No. of Totems	Generator	Boiler	Totem fuel + maint. Costs £/hour	Total Boiler (Gas) + Generator Fuel Costs £/hour	Total Capital Cost of Boiler + Generator £/hour
1	Andrews HR2	Kingfisher 125	0.60	0.99	3,130
2	Andrews CAG40	Avon 1/8	1.21	2.14	4,790
3	Lister HR6	Diplomat 2/18	1.81	2.98	6,640

From cost information approximate excess capital payback periods can be calculated.

TYPICAL PAYBACK PERIOD - Totem versus conventional Boiler and Generator

Load Factor	Typical Payback Periods in Years (Number of Totems)							
	(1)	(2)	(3)	(4)	(5)	(6)	(7)	(8)
10	1.8	2.7	3.5	5.0	3.2	3.7	4.0	5.0
20	0.9	1.4	1.8	2.5	1.6	1.8	2.1	2.5
30	0.6	0.9	1.2	1.7	1.1	1.2	1.4	1.7
40	0.45	0.7	0.9	1.3	0.8	0.9	1.0	1.3
50	0.36	0.5	0.7	1.0	0.63	0.73	0.82	1.0
60	0.3	0.45	0.6	0.8	0.5	0.6	0.7	0.8
70	0.3	0.4	0.5	0.7	0.5	0.5	0.6	0.7
80	0.2	0.3	0.4	0.6	0.4	0.5	0.5	0.6
90	0.2	0.3	0.4	0.6	0.4	0.4	0.5	0.6
100	0.2	0.27	0.35	0.5	0.32	0.37	0.41	0.51

From the above table it can be seen the very advantageous payback period in this particular economic case.

In the UK the general picture of mains electricity supply is:-

England and Wales	-	no areas where mains electricity is not available
Southern Scotland	-	no areas where mains electricity is not available
In Northern Scotland and outer islands	-	there are areas without electricity supply but which are in a programme to provide electricity in the next few years
	-	there are areas which will not be connected within the next few years
	-	there are areas where diesel powered generators provide the electricity

In those areas where the Hydroelectric Board operate diesel powered generators the cost of supplying fuel is becoming a strong argument against further installations.

In 1979 the Board lost £5M on the supply of diesel generated electricity and this year's loss is budgeted at £8M. The number of consumers served by these generating stations is approximately 30,000.

Any deliveries of gas for operating Totem in the remote isles and rural communities could be extremely costly as the existing ferries may not be permitted to carry more than small quantities, and regular supplies by special boats would require a large economic load.

Totem for Peak Lopping

The Totem system, if operating in conjunction with a mains supply, would effectively reduce the maximum demand by an amount equal to the Totem output. In addition, the installation of Totem would reduce the service capacity charges made by the Area Electricity Boards.

Totem as a Standby Facility

We do not consider the independent Totem to be a standby generator. We consider it as a cogenerator of heat and power being used as such, with the additional feature of having the capability of monitoring the electricity grid and switching to an independent role should the grid fail. Also the independent Totem is capable of being used in those areas where there is no electricity grid supply.

TOTEM DISTRIBUTORS

A limited number of approved distributors are being appointed to sell, install and offer an on-going maintenance and servicing service to customers. We recommend that the customer take out a maintenance and servicing contract with the distributor. An indication of the cost is 6.25 pence per running hour, this charge covering all labour and parts excluding oil. (The actual rate per hour will depend on the location.)

TOTEM COSTS

Standard Mk I Version

Totem	£3,440.00
Control Unit	464.00
Star Delta Starter	107.20

Independent Totem

£4,971.00

TOTEM DEMONSTRATION INSTALLATIONS

Demonstration installations have been set up. The types of demonstrations have been carefully picked to give customers a clear picture of the performance of the Total Energy Module.

When introducing to the market a product which is both a new concept and is technologically innovative there is a need to educate and inform the market of the potential. We trust this paper will contribute to that task.

TOPIC B
Community Energy

Energy Use on the Island of North Ronaldsay, Orkney

David Barbour and John Twidell

Department of Applied Physics, University of Strathclyde, Scotland, UK

ABSTRACT

North Ronaldsay is a small Orkney island with a population of 120. The energy survey studied all 49 occupied houses, the communal and service facilities, and the lighthouse. The community is now entirely dependent on fossil fuel (coal 30%, diesel 28%, heating oil 18%, petrol 15%, kerosene 5%, bottled gas 4%) for all purposes (space heat 49%, vehicles 26%, electric generation 15%, water heating 6%, cooking 4%). The lighthouse consumes diesel fuel for generation equivalent to 13% of the total island energy requirements. Per capita domestic energy consumption is 48 GJ y^{-1} (1.5 kW) compared with 29 GJ y^{-1} for the U.K. average. Household electric generators have low efficiency (10% to 20%) and average waste generator heat is greater than average space heat requirement. Fuel costs rose by 25% in 1979, well above retail price index of 10%.

The greatest energy requirements on the island are for space heat and transport. The greatest energy source is wind.

KEYWORDS

Energy; island; community; domestic; generator; costs.

INTRODUCTION

North Ronaldsay is the most northerly island in Orkney. Energy supply has always been recognised as a limiting factor on the island's development (Scott 1968, Tulloch 1974) and this is especially so now all commercial energy is imported directly as fossil fuels. There is no grid supply of electricity on or to the island, although the mainland Orkney grid is extended to all other main islands of the group and reaches the opposite coastline on the neighbouring island of Sanday. Discussions are taking place about the future energy supplies for the island, and our study was conducted to form a quantitative base for planning. With the support of the 120 strong community, we were able to visit every energy consuming house and facility, and record the amount of fuel used during 1979. The totals were checked against the records of fuel imported by ship. The survey produced a clear indication of the range of fuels and consumption, the expectations of the people, the associated costs and increases through inflation, the inefficiencies in fuel use and a guide to the most sensible strategy for future development. In particular, we as investigators, appreciated the need for energy developments to blend with the way of life on the island and reinforce employment opportunities. This paper summarises the preliminary results of our investigation.

NORTH RONALDSAY

The island lies low and exposed in the turbulent environment between the Atlantic Ocean and the North Sea, at 59° 22' North and 2° 24' West. It is 5 km long in a northerly direction, and varies in width between 3 km and 0.4 km. The coastline is rocky with some sandy bays, but there is no natural harbour. The nearest island is Sanday across the 4 km wide tidal race of the North Ronaldsay Firth. The highest point is only 23 m above sea level, and the low fields and crofts are swept by the direct force of the prevailing westerly Atlantic winds. There are virtually no trees, and only the stone dykes and buildings break the landscape. Around the perimeter there is a continuous dyke to constrain the island's unique sheep to their main diet of seaweed.

The climate is typified by partially overcast skies (average 3.3 hours of direct sunshine per day), strong winds (average velocity 7 ms^{-1}, storm gusts to 210 ms^{-1}, calm periods 10 days per year), low moderated temperature (seldom freezing, mean annual temperature 7.8°C, average maximum summer temperature about 15°C), and an adequate supply of driving rain (950 mm y^{-1}). Deep falls of snow are infrequent, winters are windy and wet rather than cold, and summers are mild with cold nights.

The land is totally divided into crofts, each associated with about 8 ha of arable land, and an estate farm of 40 ha. There is little unused land and only small areas of rough grazing. Out of 68 named crofts (Ordinance Survey 1974) and 9 other dwellings, 28 crofts were unoccupied in 1979 (30 in 1980) and 4 dwellings were used only part of the year, leaving 45 fully occupied houses. The land associated with the unoccupied crofts and those of the elderly is worked by relations and neighbours. The predominant use of the land is now for raising beef cattle. The local sheep are a small athletic variety that mainly graze on the seaweed (laminaria) exposed at low tide or driven up in bad weather. The intertidal zone can carry about the same population of sheep as the land, and so the sheep's ability to digest seaweed effectively doubles the arable land area. In summer the sheep may also graze on the central land with 'mainland' sheep brought over for summer grazing. Oats and barley are grown for animal feed after local threshing. No grain is produced now for human consumption as in previous times when there was hydro, wind and engine milling. The ratio of planted land to enclosed grazing in the crofts is only about 1 to 15.

The typical croft house is a long low single storey building with cattle byres at one end and living rooms at the other. Traditionally covered with large flagstones sealed with pitch or cement, many of the houses now have asbestos, galvanised or felt tiled roofs. Floors are traditionally stone slabs, and walls are of stone with varying degrees of cement binding. The ceiling is wood board with an attic crawl space to the sloping roof cover. In practice, no one house has all the characteristics of the typical house, and each building has a unique history after many changes and improvements. The younger families especially live in houses with few traditional aspects and many modern adaptations.

In addition to the single-storey croft houses there are six two-storey houses of standard design. These include the School house, the Church of Scotland Manse, the Health Service house for the doctor, and the manorial residence of the Laird. A major feature of the island is the Northern Lighthouse Board's complex of lighthouse and accommodation for 3 to 4 families. There are two churches in periodic use, the school with a single classroom, two shops, one community hall (another is under construction), a coastguard post for observation only, an unmanned G.P.O. telecommunications facility, a water pumping station, a pier, and an airstrip with office. There are no industrial businesses or regular commercial workshops apart from the houses and individual farm facilities.

ENERGY SURVEY

We visited every household on the island and had an extensive interview with each householder or manager. In only three cases involving small crofts did we have to rely on information by a relative. We also visited all the non-residential facilities and energy consuming installations. Our method was to explain the purpose of the survey and discuss energy matters. We asked questions referring to the calendar year 1979 from a prepared questionnaire which we ourselves completed. Questions were arranged to provide data from more than one line of reasoning - for instance tractor fuel supply was determined by asking how much was used in total, and then later separate questions determined the number of tractors, their period of use and their fuel consumption rate. Wherever possible we recorded details from fuel bills and accounts.

During the visits we made notes of building design and structure, fuel supplies, household population, historical and future aspects of energy supply, and other related matters. We did not question financial matters other than unit costs of fuels, and we promised not to identify individual households with any published results. At least four households had kept careful and detailed records, and these confirmed the reliability of the general results of the complete survey.

We also obtained details of fuel supplied to the island by outside agencies. There were 8 such suppliers, mostly based in Kirkwall. Most fuel was handled by 4 agencies on the island, although occasionally individuals imported directly. Complications also arose because the same fuels could be called by different names and different fuels by the same name. The school was supplied directly by one supplier, and the lighthouse supplies were arranged completely independently by the Northern Lighthouse Board. In general therefore we had 3 accounting 'nodes' at which to sum energy supplies - the outside suppliers, the island agencies and the individual consumers. The average flow of each fuel supply should be equal at each node, and this provided redundancy in our investigation to check the accuracy of our results. However with only one year of investigation storage of fuel and different accounting periods introduced discrepancies.

No electric generators had been metered to measure the energy supplied, and so we had to estimate electrical energy from the period of use and loading in each system. At two houses we had our own kilowatt hour meters installed, and these gave accurate readings to confirm the estimated data and measure the generating efficiency.

Data from each house, together with a photograph, was reduced to a common form by quantity of fuel and energy content. Energy flow and block diagrams were then drawn with indications of energy wastages and secondary uses.

ENERGY USE

It is important to understand energy supply in historic terms. Changes have occurred in the past, and will occur again in the future. On North Ronaldsay these changes have had clear developmental consequences and future changes will also have marked effects. Planning should assure that the changes are beneficial and not harmful.

A variety of historic energy supplies are now abandoned. The island has been populated from prehistoric times when dried animal dung and seaweed tangles were the main heat supplies until the 18th and 19th centuries. Some use of dried tangles remained until about 1910, and cow pats were still used in the 1930s. The island is considered never to have had indigenous supplies of peat or wood. Driftwood has always been valued, especially as a constructional material. Today driftwood will be used to start fires, but is virtually unknown for prolonged use. In the late 17th century and for the next 200 years peats were brought from Eday to North Ronaldsay, but the most significant change was the start of the annual coal deliveries in 1876 that have been maintained ever since. Oxen, horses and occasionally a milking cow were used as draught animals for haulage and ploughing. Animal-driven vertical-drive machines were used for threshing at some crofts. In 1771 there were 249 horses (Scott) and these remained important until the 1950s. In 1905 there were 7 herring boats based at the island, and every croft fully or partly-owned a boat for near-shore fishing. These would be powered by sail and rowing. Today there is some lobster catching, but sailing and fishing have dropped out of the island economy so that not one boat is kept at the pier. A low head water wheel powered a commercial mill until about 1900, a post wind-mill operated until 1905 (the last recorded post mill in Scotland), and an engine mill worked commercially until 1965. A wind water pump operated at the light-house until about 1952; this was removed to a croft and finally deceased in 1972. A number of wind generators were purchased or made from scrap between the 1930s and 1950s. These were used for rudimentary lighting and battery charging. Thus only coal remains of the 'historic' supplies, with all other eclipsed by oil and bottled gas. Even the unexpected and bountiful supply of carbide from the 1915 wreck has now been exhausted, although we were shown carbide lamps purchased to use that finite source and now kept in case of future provision.

There is only one purpose built solar 'glass house' for horticultural use, although a very few houses combine a South facing entrance porch with plant growing. There is no obvious use of passive solar energy by way of building design features. No housing has been thermally insulated by retro-fitting, and ceiling insulation has been incorporated in only two recent room extensions. There are 49 working diesel generators on the island, and not one is used in any form of combined heat and power mode, or had any control equipment to improve efficiency. Some householders have changed pulley diameters to improve efficiency.

Thus the island is an isolated system where all significant energy supplies are imported as fossil fuels and where there has been a minimum of energy efficiency measures. The primary supplies used are oils and petroleum, bottled gases and coal. These are used for heating, lighting and cooking; for farm machinery and tractors; for local transport; for the lighthouse, water pumping and G.P.O. stations; for electricity generation from small diesel sets; and for repair and constructional activities.

The manner of energy use in the houses varies in detail, but there is a common pattern. Nearly all houses had a form of coal or oil burning range placed in a communal room. This provides space heating and a facility for cooking, and warming food and drink. Twenty-one out of the 39 ranges we inspected were fitted with boilers for hot water supply. A few of the water heating ranges also supplied at least one central heating radiator. Without exception all the ranges were used every day throughout the year. The fire or flame would normally be lit about 7 to 8 a.m. and maintained until about 11 p.m. Our personal impression was that the heating from the ranges produced ample warmth throughout the year for the communal room. In summer kitchen temperatures varied between about 22°C and 30°C when the outside air temperature was about 14°C, and in winter kitchen temperatures would commonly be about 18°C to 20°C when outside temperatures were about 4°C. Lack of control commonly leads to excessive inside temperatures on warm days.

Most houses, but not all, had cooking rings or stoves with ovens fueled by bottled propane or butane. Heat from the cooking provided an important 'free gain' for space heat. Bottled gas was also used for space heating in free standing units, and occasionally for water heating and lighting. In addition to the kitchen range, most houses had open fireplaces in other rooms where an occasional coal fire would be lit. We saw no enclosed combustion stoves apart from the kitchen ranges.

There were 49 diesel electric generators on the island with a total capacity of 212 kW. Of these 44 generators with a total capacity of 160 kW were at independent houses, ranging from 1.5 kW to 10 kW capacity, with an average of 3.6 kW. Almost without exception the sets were made by Lister Ltd (a subsidiary of British Aerospace). Most sets were air cooled with battery starters. The main use of the electricity was for lighting, although deep freezers and other advanced devices were important in some houses. Television was common, usually operated from the household mains supply but there were a few battery sets. Households had a variety of other electrical devices, from electric tools to electric blankets. The common mode of operation for the generators was during the hours of darkness and wakefullness only (about 4 p.m. to midnight in winter, 8 p.m. to midnight in autumn and spring, 10 p.m. to midnight in summer). If the generator was on because of lighting, then the deep freeze and television would also be used, but not commonly otherwise. Thus to save generator fuel, the deep freeze was often not used in summer. Some generators are therefore run intermittently in the daylight period for devices, but every household runs its generator during darkness for lighting. Once on, a generator will be presented with an electrical load of lights and some devices, with no regard for 'conservation'. Care is taken to avoid overloading by adopting a household switching routine gained from experience. No house had any system for measuring, recording or optimizing the electricity supply. The lighthouse, was the only facility to maintain continuous electrical power, and the G.P.O. station worked on an automatic periodic routine.

The main type of agricultural machines are tractors used for general farm work and local transport. There are some stationary engines for threshing and private water supply. Cars, vans and a few motorcycles are used for journeys on the island.

RESULTS

We accounted for 7 types of fuel: coal, heating oil, paraffin, diesel oil, petrol, propane and butane. These names are as used on the island and some confusion could occur. Paraffin (also called kerosene) was as sold for lamps, and was also used occasionally in internal combustion engines and for kitchen range fuel. Some stocks of 'vaporising oil' had been used for stationary engines and kitchen ranges, but supplies had stopped. Diesel oil was as used in the electrical generators and for tractors. There was one mechanical digger using 'derv' diesel. Petrol for spark combustion engines (mostly cars) was sold from one pump and was also supplied in 50 gallon barrels. Coal arrived by ship once per year and was delivered through the island agency.

Accounting totals were made for each fuel type at the three nodal points, i.e. for the 49 occupied crofts and facilities, for the 4 island agencies, and for the 8 outside suppliers. The results are shown in tables 1,2,3,4. Deliveries were made at various times in the year, and all fuels could be stored for gradual use. Much of the consumption data was estimated by the users, and we had to rely on the information given by the suppliers without inspecting their records ourselves. Thus we would not expect the totals of each fuel used to be the same at each node. Agreement within about ± 20% would be the best expected, and by this criterion we are satisfied by the accuracy of our three sets of independent results.

Some flows of otherwise wasted energy can be useful, e.g. space heat from cooking. We have called these "free-gains". The main effect of these is on space heat and examples are usually obvious. Thus in our survey all cooking and domestic lighting become free-gains for space heat, but no waste generator heat was channelled to be a free-gain. Combustion energies of fuels and the thermal efficiencies of devices were taken from Boustead and Hancock (1978) and Leach (1979). Energy flow diagrams were drawn for each house or facility, and an example is given in Fig. 1.

Information for the energy flow diagrams was obtained in various ways. Arrows on the diagram point from the source of information. Thus electrical energy to a load could be measured by a kilowatt hour meter and the information arrow would be drawn in the direction of energy flow. Without meters, the electrical energy had to be calculated from the load usage, in which case the information arrow would be drawn contrary to the energy flow. Ideally information would be obtained in both directions simultaneously.

The total flows of commercial energy through the island system are portrayed in Fig. 2 and Fig. 3.

From the beginning to the end of 1979 the average fuel bill rose by about 25%. This rise occurred despite shipping costs being covered in September 1979 and onwards by the Orkney Islands' Council. This may be compared with the UK retail price index increase of 16%, the UK fuel and light retail price index increase of 18%, the increase in Scottish beef wholesale price of about 15%, and the increase in old age pension of 19%.

The efficiency and costs of the individual electric generators is of definite interest. We define efficiency as "ratio of electrical energy produced to fuel energy consumed", and values for a number of generators were calculated. In all cases we calculated electrical energy from the pattern of load usage, and in two of these cases we checked by installing kilowatt hour meters and measuring fuel use on a daily basis. Some household generators operate at about 10% efficiency due to the erratic operation and loading. At best generator efficiency appeared to be 20%. The average cost of fuel for the electricity at 10% efficiency is 19 pence per kilowatt hour. If the capital depreciation and repairs of the generator are added, this rises to about 24 kWh^{-1}. Inclusion of labour costs would increase the cost to about 28 p kWh^{-1} at standard rates.

On average the generator waste heat was 10% greater than the household space heat requirement, and the attraction of combined heat and power operation is obvious.

CONCLUSIONS

North Ronaldsay is now entirely dependent on imported fossil fuels for all systems requiring energy. This contrasts with the use of renewable supplies in past times, although coal has remained an important fuel for over 120 years. The per capita consumption of commercial fuel for domestic purposes (not including transport) is 48 GJ y^{-1} (1.5 kW) over 120 people. This compares with 29 GJ y^{-1} (0.9kW) for the UK average. The per capita consumption of fuel for agricultural and local transport on the island is 17 GJ y^{-1} (0.5kW).

The major requirement for energy is space heat (49% of island energy not including the lighthouse) and vehicles (26%). Space heat requirement is large because of the cold and windy climate, year round heating and lack of thermal insulation. The diesel fuel for household generators represents about 15% of the primary energy. However the electricity produced represents 3% of total end use energy because of the inefficiencies of generation. The obvious energy improvements would be thermal insulation, combined heat and power from generators, and use of the wind as a power and heat source.

ACKNOWLEDGEMENT

We would like to thank the people of North Ronaldsay for their kindness and cooperation regarding this study, and the Carnegie Trust for Scotland for financial support.

REFERENCES

Boustead, I. and G.F. Hancock (1978). Handbook of Industrial Energy Analysis. published by Ellis Horwood.

Leach, G. (1979). A low energy strategy for the United Kingdom. Science Review Ltd., London, pp259.

Ordinance Survey (1974). Ordinance Survey, Southampton. Sheets HY 75 (NE,SW,SE).

Scott, M.A. (1968). Island Saga. Alex Reid and Son, Aberdeen, pp213.

Tulloch, P.A. (1974). A window on North Ronaldsay. Kirkwall Press, Kirkwall, Orkney, pp195.

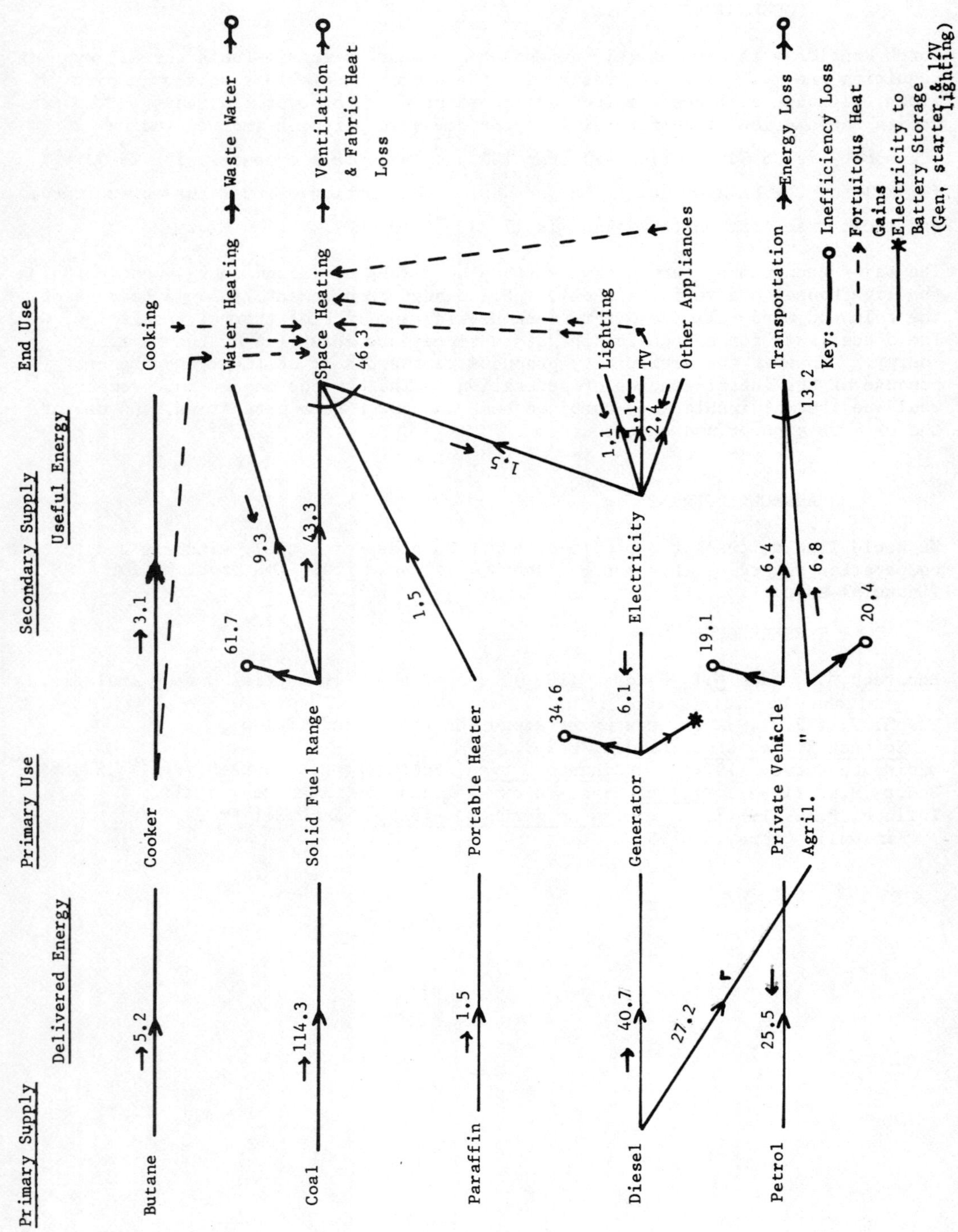

Fig. 1. **Example of household energy flow**
Unit of GJ y^{-1} (= 32W)

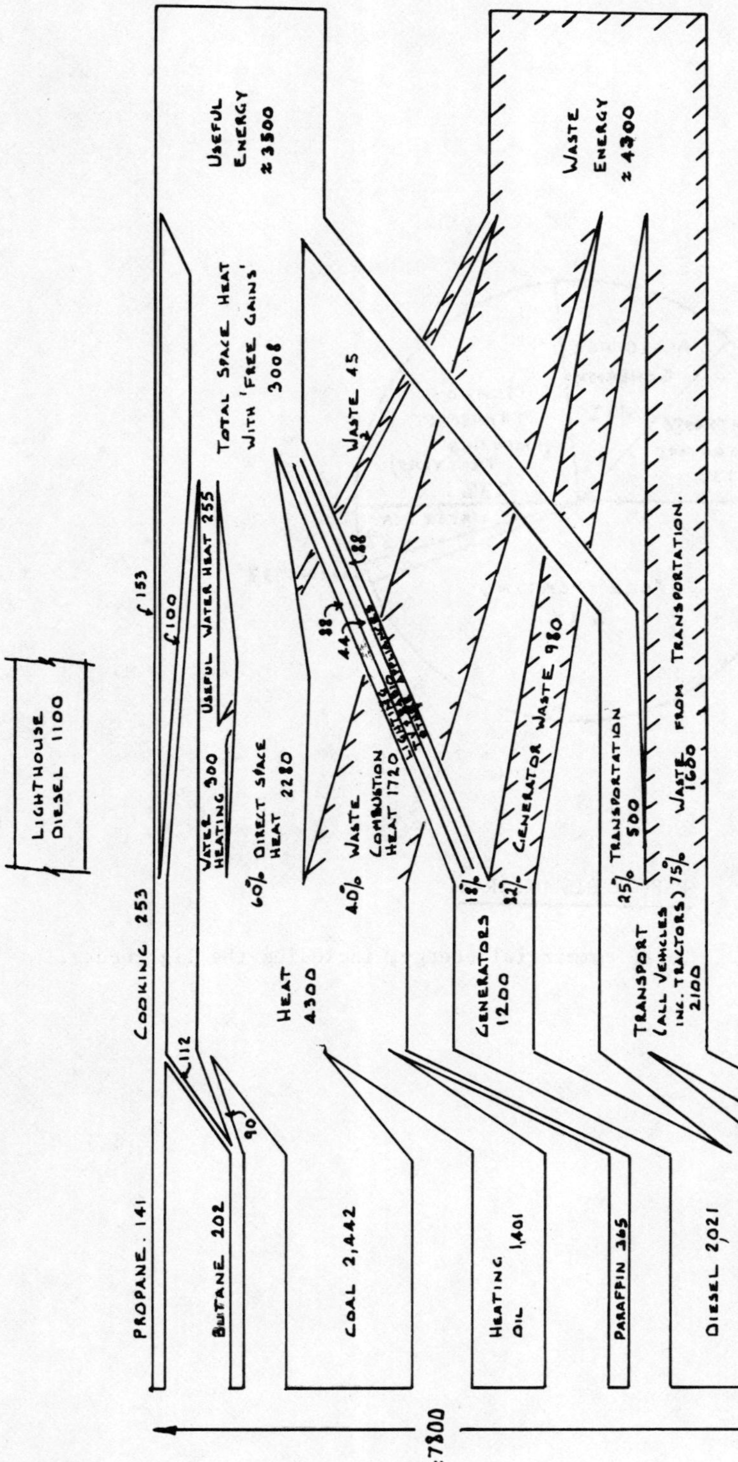

Fig 2 Energy flow diagram for North Ronaldsay : unit of GJ y^{-1} = 32 W
(Lighthouse generator fuel not included in main diagram)

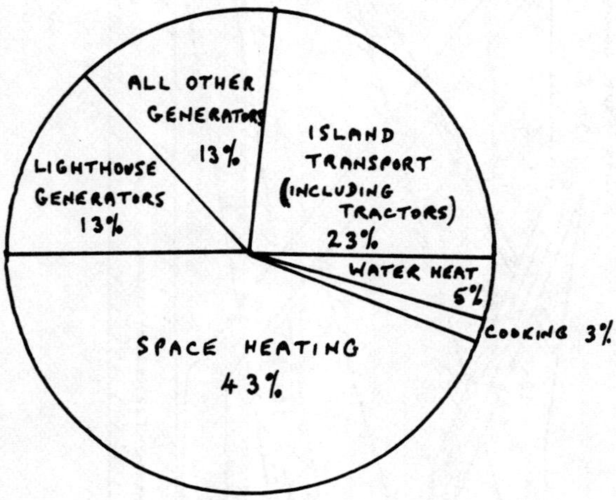

Fig 3 ENERGY PIE DIAGRAM

Total commercial energy, including the lighthouse.

	COAL/kg	HEATING OIL/LITRE	PARAFFIN/LITRE	DIESEL/LITRE	PETROL/LITRE	DERV/LITRE	BOTTLED GAS/kg
(1) MAINLAND SUPPLIERS	88,900	39,600	9,430	49,900	40,300	400	6,500
(2) ISLAND SUPPLIERS	88,000	23,600	11,900	48,600	30,500	400	7,317
(3) ISLAND CONSUMPTION ESTIMATES	84,700	40,700	8,760	62,200	29,100	400	6,950
AVERAGE OF (1),(2),(3).	87,200	34,600	10,000	53,600	33,000	400	6,920
(4) LIGHTHOUSE SUPPLY (Independent supply, additional to above)	-	-	-	29,600	1,970	-	-

TABLE 1

Primary fuel supplies to North Ronaldsay

USE	ENERGY TOTAL per year	ENERGY PER PERSON PER YEAR	
		NORTH RONALDSAY on the island	UK AVERAGE
(1) PRIMARY FUEL	7777 GJ	64.8 GJ excluding lighthouse	120 GJ
(2) TRANSPORT FUEL	2069 GJ	17.2 GJ	27 GJ
(3) DOMESTIC FUEL [(3)=(1)-(2)]	5708 GJ	47.6 GJ	29 GJ
(4) LIGHTHOUSE DIESEL AND PETROL	1186 GJ		

Table 2

ENERGY USE ON NORTH RONALDSAY
(transport on the island and domestic use)

AND UK AVERAGE
(all energy including industry, all transport, and domestic use)

Fuel	Amount used per year (from table 1)	Energy per unit	Energy used per year	Energy per person per year
Coal	87,200 kg	28.0MJ kg^{-1}	2442 GJ	20.4 GJ
Heating Oil	34,600 litre	40.5MJ ℓ^{-1}	1401 GJ	11.7 GJ
Paraffin (Kerosine)	10,000 litre	36.5MJ ℓ^{-1}	365 GJ	3.0 GJ
Diesel	53,600 litre	37.7MJ ℓ^{-1}	2021 GJ	16.8 GJ
Petrol	33,000 litre	36.0MJ ℓ^{-1}	1188 GJ	9.9 GJ
Derv	400 litre	39 MJ ℓ^{-1}	16 GJ	0.1 GJ
Bottled gas	6,920 kg	49.7MJkg^{-1}	344 GJ	2.9 GJ
TOTAL			7777 GJ	64.8 GJ

Table 3

Primary Energy use on North Ronaldsay

(excluding lighthouse supply of diesel oil and petrol)

Assuming a steady total population of 120 adults and children.

	Amount used per year	Energy per unit	Energy used per year	Energy used per person per year
Diesel for transport (40% of Primary)	21,440 litre	37.7MJ ℓ^{-1}	808 GJ	6.73 GJ
Petrol for transport (100% of Primary)	33,000 litre	36.0MJ ℓ^{-1}	1188 GJ	9.90 GJ
Paraffin for transport (20% of Primary)	2,000 litre	36.5MJ ℓ^{-1}	73 GJ	0.61 GJ
TOTAL			2069 GJ	17.2 GJ

Table 4

Transport fuel on North Ronaldsay

Assuming 40% of diesel use for transport and agricultural machinery, 60% for electric generators.

Co-operative Electricity Supply—Colonsay

Alasdair L. Munro

Development Officer
Highlands and Islands Development Board, Lochgilphead, Argyll
Director, Colonsay Community Electricity Association Ltd.[1]

ABSTRACT

This paper is an account of the setting up and operation of a community electricity supply on the island of Colonsay. This scheme has been beset by problems since its inauguration. In August 1980, however, it was announced that it is hoped that a public electricity supply will be provided in 1983 and that the Highlands and Islands Development Board would provide an operating subsidy until then to enable charges to be reduced to those levied elsewhere by the North of Scotland Hydro-Electric Board. This paper does not concentrate on the technical aspects of the scheme, but gives a broader account of the wide ranging issues involved in the establishment and running of the scheme. Readers should appreciate that, owing to the writer's own role as a director of the Association since mid-1978, some aspects of the paper may not be wholly objective.

KEYWORDS

Island development; electricity supply; co-operative company; wind-energy; hydrogen fuel.

INTRODUCTION

Colonsay is a small island in the inner Hebrides of Western Scotland. Many of the Hebridean islands are grouped closely together or lie near to the mainland shore; Colonsay is not one of these, being 30kms from the mainland and 8kms from Islay, the nearest island. This isolation is the cause of several problems of which the greatest is probably the difficulty in providing a good electricity supply.

The history of the supply of electricity to Colonsay is long and complex; it involves other important factors, as well as the island's isolation, of which the most important are probably the low population level and the role of its land-owners.

---

1 Any views in this paper are those of the author and they do not necessarily represent those of the Highlands and Islands Development Board or the Colonsay Community Electricity Association Ltd.

Colonsay is a small island measuring 11kms by 5 kms not including the even smaller adjacent island of Oransay. The population in 1980 is about 130, but like most Scottish islands, this is but a fraction of the peak population which for Colonsay and Oransay was recorded as 837 in 1851. Between 1951 and 1971 the population fell steadily from 233 to 139. Since then there has been considerable fluctuation between about 110 and 139 of whom all but one are now resident on Colonsay.

For many years, most of the land on Colonsay has been owned by one family. The father of the present owner took a paternal approach to the community in that he provided many facilities including jobs, houses and electricity. Whilst this policy contributed considerably to maintaining the island population and its services, it did not establish a secure community and further decline set in over the last 15 years or so because the present owner was unable and unwilling to maintain the same level of support. The electricity supply was one of the facilities which began to deteriorate and this led to the chain of events which resulted in the establishment of a community owned generation and distribution system.

SETTLEMENT PATTERN

Figure 1 shows the settlement pattern on Colonsay. There are three main groupings of houses. The main one on the east of the island is at Scalasaig and Glassard. Here there are some 20 permanently occupied houses, several holiday cottages, a church and the island's hotel, shop, surgery, pier and telephone exchange. The second grouping is at Kilchattan on the west of the island; this is a farming/crofting area with 13 occupied houses, a church and the island's school as well as several holiday cottages. The third grouping at Kiloran is centred on Colonsay House, the home of the island's proprietor. At Kiloran there are seven permanently occupied houses, the island's community hall plus a number of holiday cottages and flats. Outwith these three settlements, there are a further 8 permanently occupied houses scattered round Colonsay - mainly on farms.

ELECTRICITY SUPPLY BACKGROUND

The original electricity supply was installed in the nineteen-fifties by the island's owner. It provided a basic electricity supply at 110 volts D.C. from 4 generating sheds at Glassard, Port Mor, Kilchattan and Kiloran. These served most of the three main settlement groups. Over the subsequent years, improvements were carried out to the schemes and all, except Port Mor, were eventually converted to 230 volt single-phase A.C. operation. The Glassard part of the system was handed over by the Estate to the community of Scalasaig and Glassard. It was the need by 1975 for improvements to the Glassard scheme which led to the establishment in 1977 of the Colonsay Community Electricity Association Ltd scheme.

An approach was made by the Scalasaig and Glassard scheme committee to the Highlands and Islands Development Board (HIDB) for financial assistance to modernise their scheme. The Board decided, however, that rather than help with only a part of the overall system, there should be an examination of alternative ways of providing an adequate supply of electricity to the island.

PLANNING AND INSTALLATION OF SCHEME

The examination was carried out early in 1976 by Army engineers who produced a comprehensive plan for modernising and extending the whole system. This was to have been a single system based on a new generating shed at Scalasaig where a 100kW and a 50kW generators would have been installed. The main distribution was to be at 11,000 volts or 6,000 volts and would have linked up the existing distribution

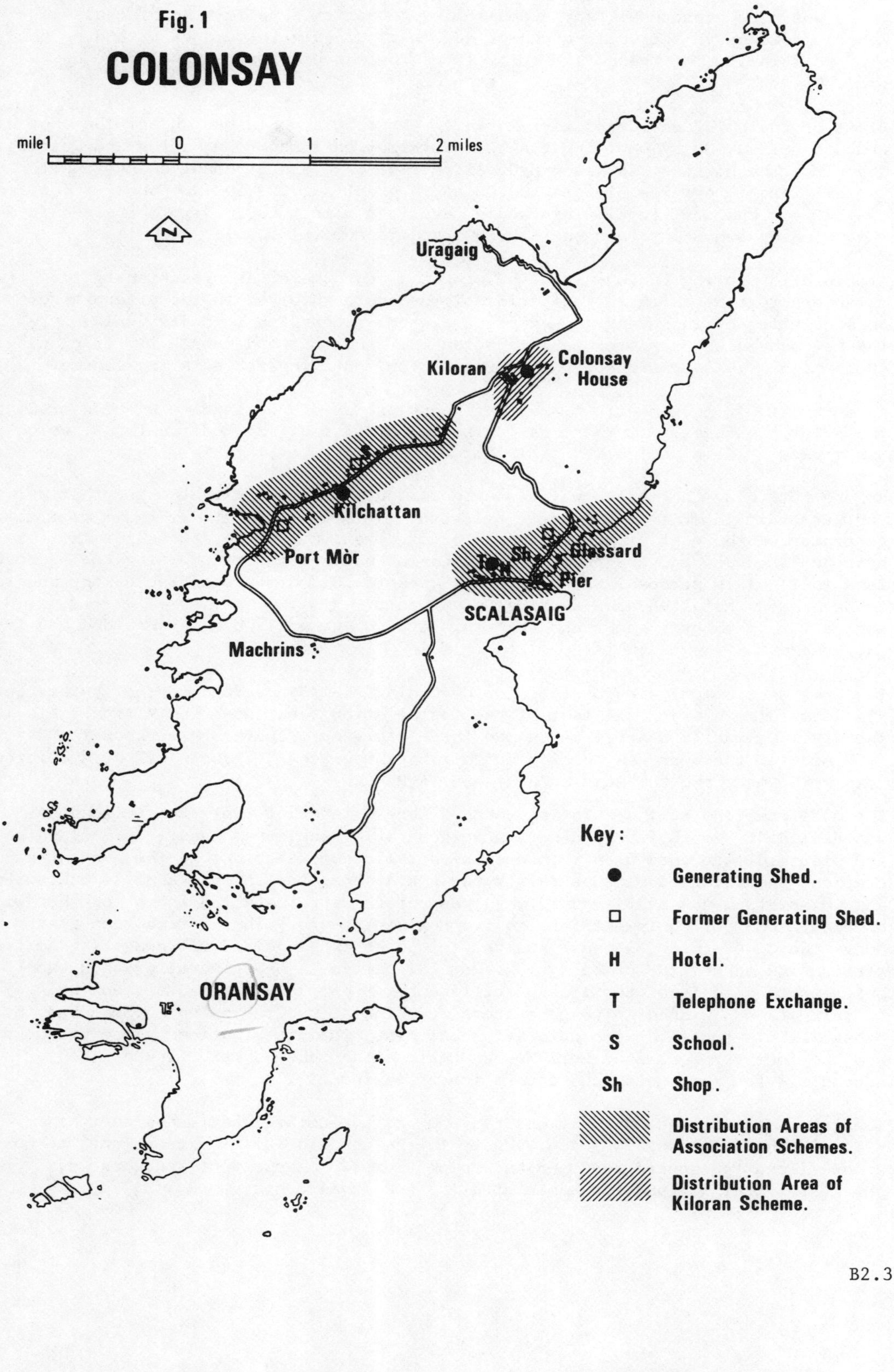
Fig. 1
COLONSAY
mile 1
0
1
2 miles
N
Uragaig
Kiloran
Colonsay House
Kilchattan
Port Mòr
Glassard
Sh
Pier
SCALASAIG
Machrins
ORANSAY
Key:
Generating Shed.
Former Generating Shed.
H Hotel.
T Telephone Exchange.
S School.
Sh Shop.
Distribution Areas of Association Schemes.
Distribution Area of Kiloran Scheme.

lines and would have extended the supply to Machrins and Uragaig, leaving out only three inhabited houses at the north and south ends of Colonsay.

There was also a proposal for a small wind generator supplemented by a diesel generator to serve the small group of houses at Uragaig instead of extending the high voltage line to there. The cost of the whole scheme was estimated to be £120,000.

Staff of the HIDB considered ways in which the scheme could be funded. The cost was reduced to £90,000 by omitting the two distribution lines to Machrins and Uragaig. The HIDB, the North of Scotland Hydro-Electric Board and Strathclyde Regional Council were to provide almost £38,000 and a grant of £22,500 was applied for from the EEC Agricultural fund. This would have left almost £30,000 to be raised from the Colonsay community in cash (£13,500) and labour (£16,000).

At a public meeting in June 1976 the proposed new scheme was presented to the islanders together with an indication of the costs involved in the alternative possibilities of improving the existing system or changing to individual generators. The new scheme was rejected largely because of the cost and complexity of running such a scheme. A preference was expressed for improving the existing scheme.

As a result of this meeting, Army engineers produced a revised plan for upgrading the existing scheme. It was this revised plan, with slight modifications, which was implemented in the summer of 1977.

Before outlining the system which was installed, it is worth mentioning that a further radical modification was considered. This involved the erection of a wind generator on the west of the island to electrolyse water, the storage of the hydrogen produced and the running of an engine-powered generator using the hydrogen as a fuel. This proposal was designed by Cranfield Institute of Technology and was to have been installed as a working experiment. In the event, the lack of a suitable engine to run on hydrogen meant that this type of generation had to be dropped from the scheme.

The Army was not only responsible for designing the improvement to the system, but its teams also carried out most of the installation work under the Military Aid to the Civil Community Scheme. Although there was some preparatory work during May 1977 most of the work was carried out during three weeks of June 1977 by a combined team of 55 men from the Regular Army and TAVR.

The modifications to the existing schemes were extensive. Only the Kiloran part was excluded. None of the other three generating sheds was used. In Scalasaig and Glassard, the shed used to house the hotel's generator became the generating station. The Estate donated a fairly new 50kVA generator in that shed to the scheme and in addition new 25kVA and 16.5kVA generators were installed there together with automatic control equipment for the two new generators. The distribution system was converted to 4 - wire, 3-phase and the telephone exchange and microwave transmitting station was connected to the "grid". At Kilchattan, a new generating shed was built in the middle of the linear settlement. Local volunteers assisted the Army in building this new shed. Two generators were installed, one being 16.5kVA and the other 12.5kVA together with automatic controls. The distribution system was upgraded and the formerly separated Port Mor and Kilchattan schemes were linked. (See Appendix 1 for more detailed description of equipment).

The total cost of the project was £35,000, as against the estimated cost of £22,000. The extra cost was due largely to the need to instal extra plant, to build a new Kilchattan generating shed and to provide for a supply of spare parts, engineer's training and for administrative and legal costs.

Most of the cost was provided by various public bodies - the HIDB, the North of Scotland Hydro-Electric Board, Strathclyde Regional Council and Argyll and Bute District Council.

CO-OPERATIVE STRUCTURE AND OPERATION

The Colonsay Community Electricity Association Ltd which was established to run the scheme is a company, limited by guarantee. The members of the company are the consumers. The running of the scheme is the responsibility of a board of directors most of whom are elected by the members to serve for three years at a time. In addition the HIDB, Argyll and Bute District Council and the Estate have appointed directors to the Association; these directors have the same powers as elected directors. In practice only the HIDB and Estate directors attend meetings on an occasional basis.

When the Association commenced operations, it was charged, under the terms of public assistance provided, to run the scheme in such a way that it could be self-supporting. The Association was required to set aside £1000 per annum to establish a sinking fund to cover the depreciation of equipment and the transmission system.

The Association has one employee - a part time maintenance worker. The person who was first appointed received training from the North of Scotland Hydro-Electric Board and the equipment suppliers. Initially he also worked part-time for the Estate, but this was simplified by making him a full-time employee and by the Association sub-contracting his time to the Estate. When he left Colonsay at the end of 1979 to emigrate to Canada, he was replaced by a young local man who is a physics graduate and a teacher and who has taken over a family croft. He is employed part-time and does not work for the Estate. Major overhauls and repairs are carried out by visiting engineers. The administration of the company and its operations is largely in the hands of the company secretary and treasurer who, at the time of writing, are the local doctor and social worker respectively. Both of these office-bearers have a considerable work load. Another of the directors acts as meter-reader. Local fuel delivery is arranged by the island's sole shop-keeper who is also chairman of the board of directors. Fuel is collected in Oban using the Association's own 2250 litre tanker-trailer.

When the project was being planned, it was hoped that, with wind-power, the cost per unit would be 7p and it was largely on this understanding that community support was forthcoming. In the event, without wind-power, it was clear that the operating costs would be higher. The initial tariff was therefore set at the much higher level of 15p per unit with special tariffs being arranged for the Post Office and Strathclyde Region. The east-side Scalasaig part of the scheme provides a 24-hour service, largely because of the contractual arrangement with the Post Office. At Kilchattan, on the west side of the island, a restricted service is operated; initially electricity was supplied in the morning and evening, but eventually, partly because of the school's needs, it was supplied for a continuous period throughout the day.

Technically, there have been no major problems since the scheme started. The overall consumption of electricity is broadly in accord with the original estimate prepared by the Army. The initial consumption was higher than planned, but there was a reduction in 1978 when the hotel, a major consumer, changed hands. Apart from that most domestic consumption is at a minimal level with little scope for much reduction.

Between April 1979 and April 1980, 64,700 units were consumed of which only 13% were attributed to the west-side of Kilchattan scheme. By far the largest user was the Post Office, taking some 26,000 units. The hotel consumed almost 12,000 units and the domestic consumers in total accounted for under 20,000 units.

The average domestic consumption was 488 units, but this ranged from 105 to 2027.

FINANCIAL AND OPERATIONAL PROBLEMS

The plant, particularly at Kilchattan, is generally running well below capacity and this has resulted in higher than planned maintenance and repair costs. Separate records are not kept of the fuel used at each scheme, but over a period of 17 months, the efficiency of the total system was 14.8% in terms of the energy value of the fuel consumed in relation to the electricity distributed.

This very high charge for electricity (15p per unit) meant that from the very start of its operations, the Association was not fulfilling its expectations in the eyes of the community. The individual consumers appeared to regard the Association, or more particularly the directors, as a remote body in which they had little part to play. As a result decisions were taken which probably contributed to the later financial problems and there emerged a reluctance to institute any changes to the operation of the scheme such as the tariff structure in case they would initially result in even greater criticism, although they may have produced significant longer term improvements.

The 15p per unit charge was most unpopular. At the beginning of 1978 the directors reduced the price to 12p in an attempt to relieve the burden and to stimulate greater consumption. This was initially intended as a short-term measure, a "special offer", but from May 1978 the tariff was confirmed at the even lower rate of 11p per unit. This rate continued to be charged until August 1979 when inflation and the worsening financial position forced it up to 14p. The 14p rate was applied for only 6 months and a further increase to 18p was implemented from January 1980.

The accounts for the year up to May 1979 superficially suggest that the Association was holding its own. From an income of £7,500, there was an operating profit of £88 (See Appendix 2 for a summary of the accounts for 1978/79 and 1979/80). The accounts, however, do not include any depreciation. Therefore the £1,000 which the Association was obliged to place in the sinking fund was financed out of working capital. This and inflation resulted in a deteriorating cash position from early 1979. In July 1979 the Association was relieved by the HIDB of its responsibility to maintain a sinking fund and the Association reluctantly converted the fund to cash knowing that this could only be a short-term relief.

The disenchantment with the scheme caused further fundamental problems. The directors found that community participation in outstanding tasks was almost impossible to organise. By the summer of 1980, three years after its erection, the rendering of the Kilchattan generating shed had not been completed and neither had storm damage to the roof which occurred late in 1979 been repaired.

From an early date in its operations, the directors felt that it would be difficult for the Association to continue to run in the way it had been set up. Indeed the first attempts to obtain a subsidy were begun as early as August 1977. These attempts built up into a campaign which resulted in the plight of the Association and of the island receiving considerable publicity. The Association, and in particular, its secretary fought very hard to obtain the promise of a subsidy and a public electricity supply and this was finally announced in Parliament in August 1980.

CONCLUSION

The proposed subsidy to be provided by the HIDB will hopefully enable the Association to continue in being until the public supply is connected in 1983. It is too early yet to make a wholly objective assessment of the history of the

Association. What is clear is that it has not been a happy period, even although most of Colonsay has been provided with a good supply of electricity. Mistakes may have been made in the technical design of the scheme in that it has probably lacked flexibility of operation. It may also be the case to note, with hindsight, that it may have been unfair on the people of Colonsay that a fairly ambitious community company was established in an island whose people had had little opportunity in the past to develop the self-reliance and independence which is often seen in many of the small islands of the Outer Hebrides, Orkney and Shetland.

ACKNOWLEDGEMENT

The author wishes to convey his thanks to those who helped in the preparation of this paper by supplying additional information. In particular, he wishes to thank David Henderson of the HIDB and Dr Andrew Hall-Gardiner of Colonsay. Hilary Wyper, University of Strathclyde provided the information from which the energy efficiency of the scheme was calculated and Dr John Twidell, University of Strathclyde collected the data for Appendix I.

REFERENCES

No published references were used in the preparation of this paper. The main sources of material were internal papers and reports of the Highlands and Islands Development Board and of the Colonsay Community Electricity Association Ltd.

APPENDIX 1

INSTALLED EQUIPMENT

(a) Scalasaig - Set 1 25kVA, 35A and 8.3kW max per phase, 415/240V, 50Hz. Lister diesel engine - No 4191 HR3 26V 32.25HP, 1500 RPM. Puma alternator, type R25L5, No 11279/1
Set 2 - 16.25kVA, 23A and 5.3kW max per phase, 415/240V, 50Hz. Lister diesel engine - No 3885 HR2A 26V, 21.5HP, 1500 RPM. Puma alternator, type R16L4, no 11283/1
Set 3 - 60kVA, 3 phase not in regular use. BMC system formerly used to provide electricity for hotel.

(b) Kilchattan - Set 1 - 16.25kVA As set 2, Scalasaig. Engine number 3942, alternator number 11278/1.
Set 2 - 12.5kVA, 17.4A and 4.2kW max per phase, 415/240V, 50Hz. Lister diesel engine - no 526 ST 27, 18HP, 1500RPM, Puma alternator, type R12 5L3, no 11277/1.

APPENDIX 2

SUMMARY OF INCOME AND EXPENDITURE ACCOUNTS

Year ending May	1979	1980
Income		
Electricity supply	£7,353	£10,004
Other	148	305
	£7,501	£10,309
Expenditure		
Wages (net cost)	£1,762	£ 2,303
Fuel oil	3,756	5,460
Repairs	973	1,411
Other	922	1,307
	£7,413	£10,481
Surplus (Deficit)	£ 88	(£ 172)

The Fair Isle Community Electricity Scheme

David Wheeler

Fair Isle Electricity Committee, Fair Isle, Shetland

ABSTRACT

Due to its isolation, Fair Isle has no public electricity supply for its 70 inhabitants. Electricity is produced for a few hours each day by diesel generators. Though the system is efficient, running costs are high and continuing to rise. Fair Isle Electricity Committee is looking for an alternative method of electricity generation. With a mean wind speed of 9.3 m/s, wind power would appear to be the solution for Fair Isle, and such a project is already in its very early stages.

KEYWORDS

Island; electricity; efficient; expensive; alternatives.

THE ISLAND

Description

Fair Isle, owned by the National Trust for Scotland, is situated approximately half way between the two island groups of Orkney and Shetland off the north east of Scotland. The climate is oceanic in nature. Summers are cool and often foggy. Though almost on the 60° latitude with a possible 18 hours daily sunshine in summer, total sunshine rarely exceeds 200 hours in the sunniest month. Winters are mild and stormy. The annual mean temperature is 7.6^{o}C, and the average wind speed 9.3 m/s. During 1979, gales were recorded on 112 days.

The island mailboat, Good Shepherd III, a 15 metre converted fishing boat, makes the 30 kilometre trip to the south end of Shetland twice a week in summer and once a week (weather permitting) in winter, carrying passengers, mail, fuel and stores. After each winter trip, the boat is hauled up on to the island slipway. In addition, a Loganair 7-seater Britten Norman Islander operates a scheduled air link with Shetland four times a week in summer and twice a week in winter, and also with Orkney once a week in summer.

Population Distribution

The resident population is about 70 with a large proportion of young people, but this increases to 100-110 during summer, when many visitors are attracted to the island. At the southern end lie the eighteen crofts and the shop, nurse's house, school and schoolhouse, hall and two churches. All these are connected to the underground electricity distribution cables from the centrally placed generators. The South Lighthouse is close to the crofting area, while the North Lighthouse is on the very northern tip of the island almost five kilometres away. Both have their own separate electricity supplies. The National Trust for Scotland bunkhouse accommodation is at present not connected to any electricity supply. It is situated between the crofting area and the South Lighthouse. It is hoped to connect it to the island electricity supply for summer use but, due to the 1200 metre cable run from the central generator, power loss means that lights only could be used. The Bird Observatory is sited some 2 kilometres NE of the crofting area, and has its own electricity supply.

COMMUNAL SUPPLY

History

On taking ownership of Fair Isle in 1954, the National Trust for Scotland embarked upon a programme of modernising, as funds permitted, croft houses that became vacant. By 1960, the provision of electricity was considered essential. With assistance from islanders and volunteer parties, three 240 volt single phase circuits, powered by 7 kW Lister generators, supplying all croft houses, were operational by 1962. Each consumer paid an annual fee to the National Trust for Scotland. Three generators (with two spare) which each supplied five to eight consumers were operated by the crofter on whose premises the generator was located. The Trust, as owners, maintained the plant, while fuel was supplied by the consumers.

By 1974, the cost of maintaining the generators was equal to the rental. Following advice from the North of Scotland Hydro Electric Board, a replacement generator plant was established in a central location on the island. A 52 kW Lister powered generator was installed in a specially modified breeze block building in the Houll quarry. The existing circuits were linked to this generator. A 20 kW Lister powered generator was later purchased as a standby and for use when the demand could not justify the operation of the larger plant.

The total cost of the 1976 reinforcement scheme was £11075. This was financed by grants from the Highlands & Islands Development Board - £4100, the North of Scotland Hydro Electric Board - £3500, and the National Trust for Scotland - £3475. On completion of the installation, all the plant, cables and buildings were leased to the Fair Isle Electricity Committee. The Committee became responsible for all operational and maintenance costs, including the provision of a £400 annual allocation to a Replacement Fund.

The Social Work Department of Shetland Islands Council has recently given money to provide a laundry service for Fair Isle's pensioners. This is to be installed in an extension to the National Trust for Scotland bunkhouse accommodation. A 6 kW generator has been purchased, to be operated by the Fair Isle Electricity Committee.

Installation

Both main generators and their control gear are housed in the same building. As well as the output from the alternators, starting, stopping and plant protection circuits are taken through changeover switches, thus avoiding the duplication of control panels. Following the changeover switch, alternator output is fed to the four fused distribution switches in the generator house. The main panel, as well as output and load monitoring facilities, also contains a hand-wound fourteen-day time switch with two on/off cycles each day for starting and stopping the selected engine. It is also possible to start or stop the engine manually. If an engine fails to start, or while running one of the three protection circuits (oil pressure, engine temperature or alternator output) detects a malfunction, the plant is automatically shut down and 'locked out' with the appropriate alarm lamp showing. The engine cannot be re-started until the fault is rectified.

Generator Details

Set 1 - 65.7 kVA 73 amps and 17.5 kW maximum per phase, 415/240 V, 50 Hz.
Lister diesel engine - no. 15HRS6A25V 80HP, 1500RPM. Brush AC Generator

Set 2 - 25.0 kVA 27.8 amps and 6.7 kW maximum per phase, 415/240V, 50 Hz.
Lister diesel engine - no. 842HR3A25V, 32.25 HP, 1500RPM. Brush AC Generator

Laundry Set - 7 kVA Single phase, 220-240V, 50 Hz.
Lister diesel engine - no. 5770ST227 - 04. 10 HP, 1500RPM. Brush AC Generator

Transmission

Transmission is by underground armoured 4-core cable to three main distribution points, each fused. The various premises are than fed from these points by 2-core armoured cable. This system allows a flexibility in balancing phase loading, given different consumer requirements. Each consumer receives a single-phase 225-240V supply through a meter. No consumer is more than 1000 metres from the generator house, but some power loss is inevitable on the longer runs. It is hoped to include voltage regulators at strategic points in order to minimise this power loss.

Operation

It is uneconomic to provide a 24-hour electricity supply. The operating times are 8.30 am - 10.30 am and 7.00 pm (or sunset when earlier) - 11.00 pm. On special occasions - dances, etc. - the evening shutdown is delayed for an hour or two.

Charges

The Committee is required to cover its costs and put £400 per annum into a Replacement Fund. To encourage greater use, electricity is sold on a sliding scale of charges, though the two churches are supplied free. For example, during the quarter ending 19th July 1980, all consumers paid a rental of £4.00, the first 100 units were charged at 8p, the next 200 at 4p, the next 200 at 2p, and further units at 1p each.

During the year 20/4/79 to 19/4/80, 39,606 kWhr of electricity were sold to the 23 consumers, the majority using between 1000 and 2000 kWhr. Only three used more than 3000 kWhr and three used less than 500 kWhr. Electricity was provided for a total of 2789 hours. During this period, each unit was sold at an average price of 6.24p, though it cost 8.77p to produce. The difference was covered by the Committee receiving a grant of £1000 from the Shetland Islands Council, through the Charitable Trust, towards the operation of the scheme. In the same period, the North of Scotland Hydro Electric Board tariff on the Shetland mainland was about 3p a unit. (See Table 1, Fig. 1 and Fig. 2).

TABLE 1 Operating Details

PERIOD	20/4/76 to 29/4/77	20/4/77 to 19/4/78	20/4/78 to 19/4/79	20/4/79 to 19/4/80	Estimate for 1980-81
kWhr used	23769	29502	37366	39606	41000
Hours generating	2736	2678	2780	2789	2750
Litres fuel used	18516	17820	17057	16452	16500
Average load kW	8.69	11.02	13.44	14.20	
kW per litre fuel	1.28	1.65	2.19	2.41	
Efficiency per cent	12.3	15.8	20.9	23.0	
Fuel cost per litre	7.93p	9.35p	9.69p	12.38p	17.0p
Total fuel cost	£1548.22	£1760.56	£1755.92	£2155.27	£3000.00
Total operating cost	£1677.83	£2582.12	£3013.37	£3472.89	£4360.00
Fuel only cost					
per kW	6.51p	5.97p	4.70p	5.44p	7.3p
per hr	56.59p	65.74p	63.16p	77.28p	109.1p
Total cost					
per kW	7.06p	8.75p	8.06p	8.77p	10.6p
per hr	61.32p	96.42p	108.39p	124.52p	158.5p

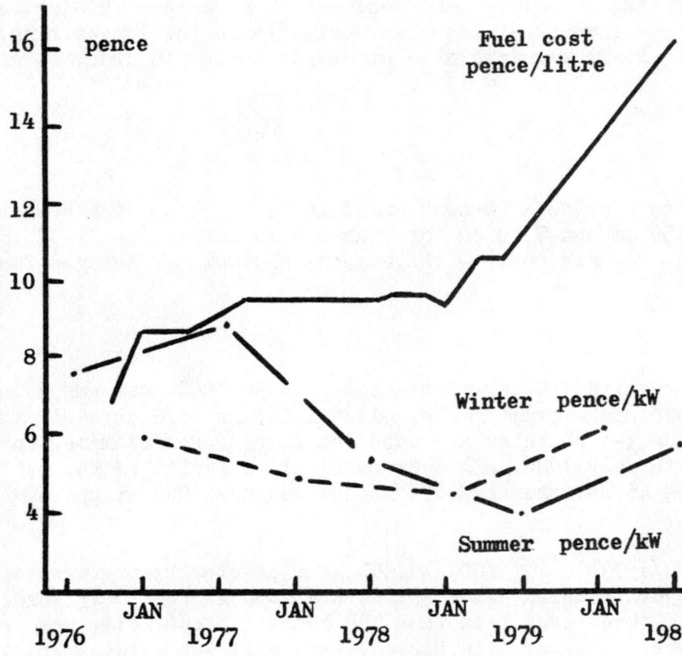

Fig. 1. Fuel and Generating Costs
(Fuel element only)

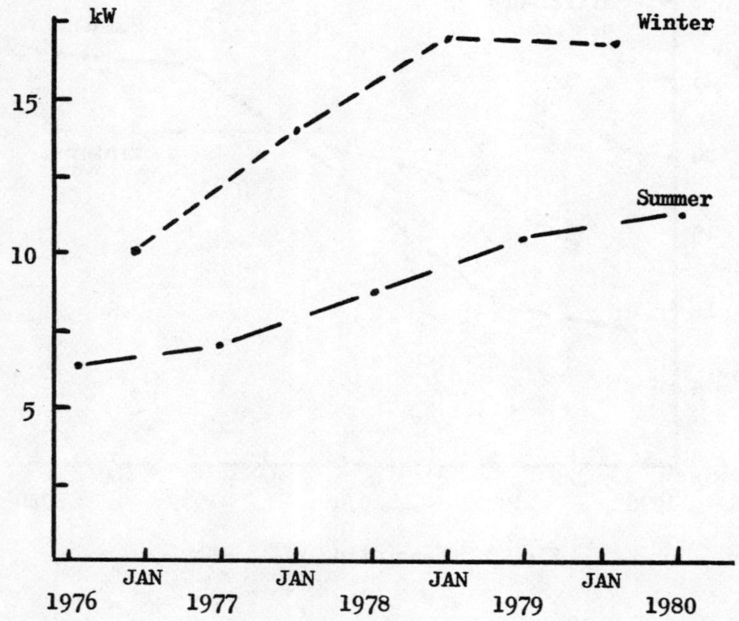

Fig. 2. Average load kW

Efficiency

Due to the limited hours of operation each day, the overall efficiency is high at 23 per cent. The present system is working well, but the 52 kW generator has proved too large for our present requirements. It had been hoped to remove the turbo-charger of this engine and so improve fuel efficiency, but this has proved impossible as it is an integral part of the engine. The 20 kW generator, though having an output nearer the island's present needs, cannot cope with peak loads except during the summer months. (See Fig. 3 - Generating efficiency).

Maintenance

Day to day maintenance is undertaken by a local engineer, who is paid on an hourly basis. A visiting engineer, with the assistance of the local engineer, carries out annual overhauls and any major repairs or maintenance.

Administration

The Fair Isle electricity supply is operated by an unpaid Committee of four, consisting of Chairman, Secretary, Treasurer and one other member. The Committee is elected at an Annual General Meeting of the Fair Isle Electricity Council - all the consumers.

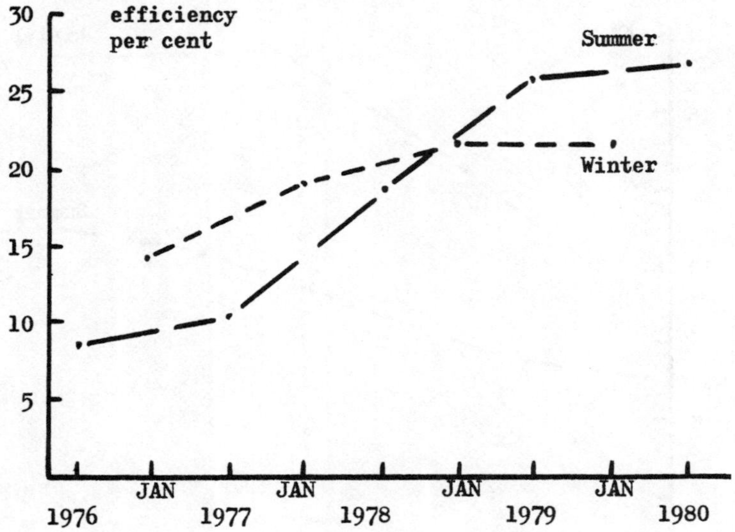

Fig. 3. Generating efficiency

PRIVATE SUPPLIES

As well as the communal production, electricity is also generated at a number of other points on the island.

Lighthouses

At both lighthouses electricity is generated for some fourteen hours each day - by a 4 kW generator at the South Lighthouse and a 6 kW generator at the North Lighthouse. The total annual consumption for both is estimated at 6000 kWhr, mostly used for domestic lighting, radios and televisions. Electricity is also used indirectly through batteries for the automatic navigation light and fog horn at the North Lighthouse. The South Lighthouse still has a paraffin navigation light and its fog horn is operated by diesel engine driven compressors.

Fair Isle Bird Observatory

The Observatory runs either a 7.5 kW or a 10.5 kW generator, depending upon requirements, for about 10 hours each day. Estimated annual consumption is 6000 kWhr, mostly used for lighting, freezer, food mixer and other domestic equipment.

Independent Broadcasting Authority/Post Office Television and Telephone Relay Services

The IBA and Post Office operate three 25 kW generators to give a 24-hour supply to power television and telephone relay services. This generating station is towards the north end of the island, some 2 kilometres from the crofting area. Though approached to provide the island population with 24-hour electricity, these bodies were unable to do so as they require a very stable supply. The total annual consumption is estimated at 150,000 kWhr, much of which is used for air conditioning electronic equipment.

Private Generators

Of the 23 communal supply consumers, 9 also have their own generators - ranging from under 2kW to 7 kW in output. Annual consumption of electricity from these generators is estimated at 3800 kWhr. An important reason for these private generators is to give deep freezers a "boost" during the infrequent warm spells on Fair Isle and to assist in freezing down foodstuffs, though it has also been found possible, with care, to operate deep freezers safely on the minimum of six hours communal electricity a day.

FUTURE DEVELOPMENT

In the face of ever increasing fuel costs - which in August 1980 were already up to almost 16p per litre compared to the average 12.4p per litre for the period 20/4/79 to 19/4/80 - the present system is going to become more and more expensive. For this year, the Fair Isle Electricity Committee is having to ask Shetland Islands Council for a grant of £1500 towards the operating costs of the scheme. Without extra grant assistance, electricity charges will have to increase. This would result in consumers economising in electricity use, another inevitable price rise, and yet lower demand. As it is unlikely that more consumers can be included in the system, other than the two Sheltered Housing units proposed for construction in a few years' time, the present consumers must be encouraged to use more electricity to keep costs down. More electricity would be used if it were cheaper.

One alternative is that Fair Isle might be linked to the North of Scotland Hydro Electric Board supply in either Orkney or Shetland. This is probably ruled out by the difficulty of safeguarding the cable and also by cost - an estimated £2 million.

With an annual mean wind speed of 9.3 m/s, the most attractive alternative would be the generation of electricity from wind power, using the present system only as a standby during periods of light winds. At the moment, the Fair Isle Electricity Committee is in the early stages of planning a medium size wind turbine generator project.

ACKNOWLEDGEMENTS

The author would like to express his gratitude to Mr Barry Sinclair and Mr Neil Thomson for their assistance in the preparation of the data for this paper. Thanks are also due to Mr Malcolm Adamson for his comments on the Fair Isle Electricity scheme, and to Mr David MacLehose of the National Trust for Scotland for information on the original electricity supply and the financing of the present installation.

Thirty Years' Experience of Private Hydro-electric Generation

Charles Crichton

Monzie Castle, Monzie, Crieff, Perthshire

ABSTRACT

The Scottish history of private and public electricity generation and supply is traced. The 200 kw Monzie Hydro-Electric Scheme was commissioned in 1950 and supplies electricity to Monzie Estate in Perthshire and additional kilowatt hours are sold to the grid. The Scheme is examined and the problem discussed.

KEYWORDS

Hydro-electric; Private Supply.

PAPER

Text of paper presented at the Conference not available for the published Proceedings.

Diesel Generation on Scottish Islands by the North of Scotland Hydro-electric Board

H. B. Macdonald

ABSTRACT

The Hydro-Electric Board assumed responsibility for electrical power supply on Scottish Islands in 1947 and the installation has now reached fairly extensive levels, comparison being provided with Scottish mainland. Planning and economics are discussed and parameters for system extension and plant margins given. Justification of diesel plant as opposed to other forms of prim mover fuel requirements and factors affecting siting noted. Operational costs are briefly commented on and future development discussed.

KEYWORDS

Diesel Generation; Scottish Islands; Plant margin; plant selection; fuel; siting; cost; future development.

INTRODUCTION

The North of Scotland Hydro-Electric Board became responsible for the generation and distribution of electrical power throughout the North Scottish mainland and off-shore islands upon the nationalisation of the industry in 1947. At that time there were no established undertakings or companies supplying the major island groups with the exception of Lerwick, Kirkwall and Stornoway, where small town council stations existed supplying only their respective townships. In addition to these stations there were a few scattered Ministry of Defence installations that had been constructed during the war but essentially there was little electrification in the islands. No High Voltage network existed in any of the major island groups.

Today the installation comprises on the major island groups served by diesel generation, ie Shetland, Orkney, the Western Isles and Tiree, some 117 MW of generation plant, 3000 km of HV distribution network and 80 km of submarine cables supplying islands within these groups. The total number of consumers connected is approximately 31,000, representing 95% of all potential consumers within these groups. Four major stations have been established at Lerwick, Kirkwall, Stornoway and Loch Carnan (South Uist) and these serve the major island groups, while a small station on Tiree serves the interconnected islands of Coll and Tiree. These stations are all isolated and are not interconnected, although consideration is being given to the interconnection of Stornoway and Loch Carnan on the Western Isles, such interconnection, while having little effect on operational costs, would, by enabling the stations to be considered as a single unit, reduce the spare plant capacity with consequent saving of capital cost.

The station capacities are given in the following table. The three major stations, Lerwick, Kirkwall and Stornoway were originally constructed in 1952 and have each been subsequently extended. The original building size was designed for an ultimate capacity of eight 1 MW units Polar KM-48 Diesel Generators of which three were originally installed in each station. These units have since been removed and replaced with 4.5 MW Mirrlees Blackstone Generators in the same space; an indication of the power increase that has been achieved by diesel engine designers over fifteen years.

Loch Carnan was constructed by the Board to replace Daliburgh Generating Station which had been outclassed by the increasing demand. Fuel had to be delivered to it by road tanker a commitment that had been increasingly difficult to meet and the load centre had shifted. Accordingly in 1972 Loch Carnan was constructed in a location where fuel could be supplied by sea.

Station	Name Plate Rating	MW Firm Capacity	No of Sets	Maximum Demand 79/80
Lerwick	41.8	29.8	11	30.8
Kirkwall	27.9	21.0	8	17.5
Stornoway	30.3	19.9	9	18.6
Loch Carnan	11.7	8.4	5	7.0
Tiree	1.8	1.26	4	1.0
Flotta (Gas Turbine)ϕ	3.3	-	1	Connected to Orkney System
Bowmore*	6.0	6.0	3	Standby

ϕ Flotta burns flare gas which is subject to interruption and is consequently not considered firm.

* Bowmore, Islay is a station maintained as a standby station primarily against failure of the mainland - Jura cable.

COMPARISON OF ISOLATED AND MAINLAND SYSTEMS

The Maximum Demand and Total System Requirement is illustrated by Figures 1 and 2 for the period 1955-1980. Similar curves for the Scottish mainland system have been superimposed for the same period. It can be seen that, while the rate of growth in the Scottish mainland system declined in the 1970s, the Isolated Systems trend was for demand to increase towards the end of the 1960s and to be further sustained in the 1970s, mainly by oil development. The System Load Factor also improved over this period from 34% to 50% compared with equivalent Annual Load Factors of 43% to 56% on the Scottish mainland system.

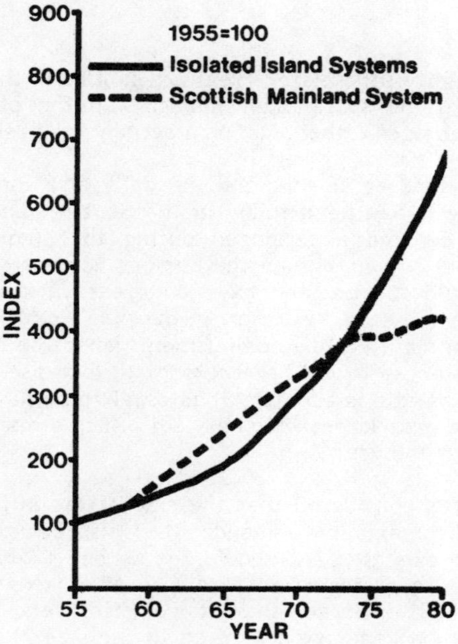

FIG 1 - MAXIMUM DEMAND

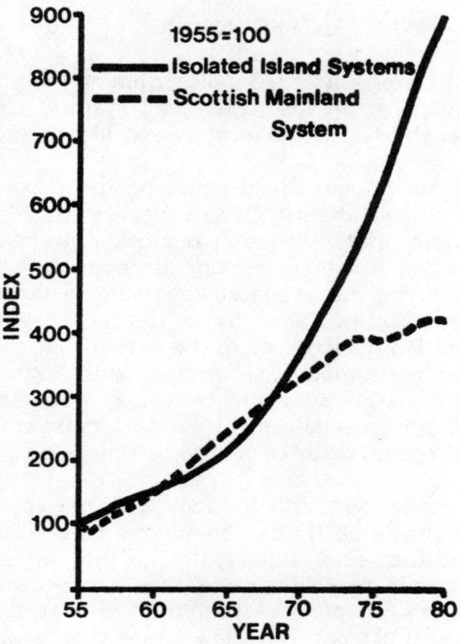

FIG 2 - UNITS SENT OUT

The Daily and Annual Load Variation experienced in the isolated areas is characteristic of that experienced in the Scottish mainland, the overnight load in winter dropping to 50% of the maximum load. There is less tendancy in winter for the load to drop significantly following the midday peak owing to shorter daylight hours. The electrical load on all islands is essentially a resistive load (lighting and heating) and system power factors are of the order 0.95-0.97.

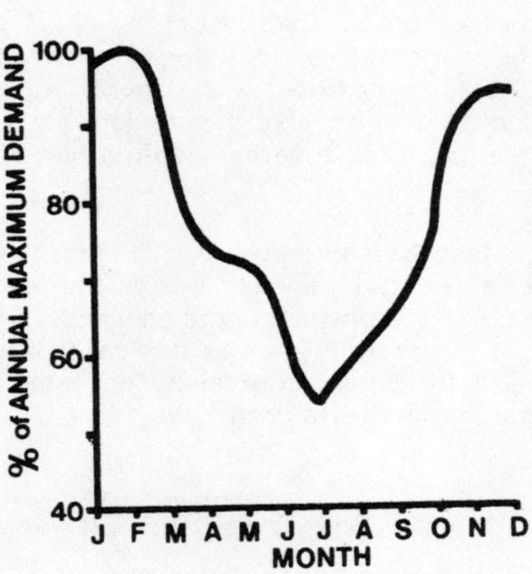

FIG 3 - ANNUAL LOAD CURVE

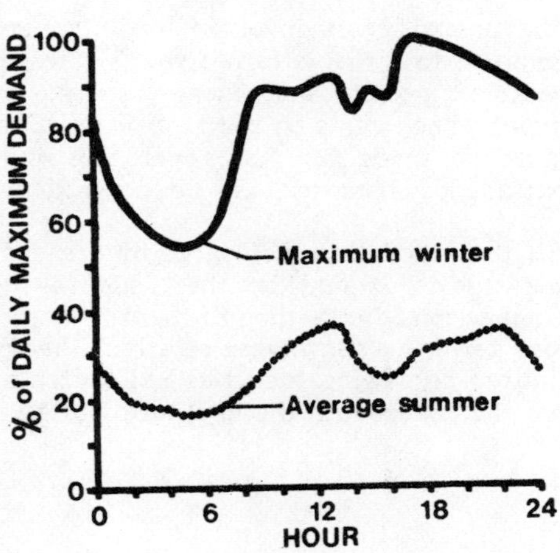

FIG 4 - DAILY LOAD CURVE

PLANNING AND ECONOMICS

The criteria adopted in determining the plant capacity for isolated areas are similar to those applied to an interconnected system, apart from the spare plant margin and the plant availability. These must be considered as finite set sizes rather than on a percentage basis.

The plant capacity depends on the maximum demand to be met and the daily and annual variations which will be experienced. Previously it had been found in the Scottish Island system that, with the reduction in maximum demand experienced during the summer months, it was acceptable to install one additional set more than the number necessary to cater for the expected maximum demand. Should the demand have no great variation, either daily or annually, a greater margin will be required. Further, if the number of sets installed is increased to the extent that the normal maintenance commitment cannot be met over the summer load period, under such circumstances it will be necessary to increase the plant margin by a further set. if a two set plant margin is adopted, it is usual to nominate the outage requirement as the largest set and the next largest capacity set, such a margin will enable maintenance to be scheduled throughout the year.

In connection with the individual set capacity, it is considered that the size of the largest set should be limited to approximately 30% of the maximum demand. It is also desirable that the set size is related to the minimum load expected and should not exceed 125% of this load to achieve economy in operation. When considering an extension to an existing station, it must be remembered that if a new unit is larger than the existing sets, the 'largest set out' criterion means that the gain in firm capacity is limited to the size of the previous largest set.

The spare capacity maintained by the Board in its diesel stations is approximately 30% of total plant.

In the case of isolated generation, as load grows, periodic reviews should be made to compare the economics of installing further generating plant with the alternative of reinforcement from the main system, either using the existing generating plant as standby plant or providing a duplicate supply to maintain the required security.

The annual costs involved must include capital charges on all investment which is not common to either alternative. In the case where the main system would be extended, the cost of power and energy charges (including losses), together with operation and maintenance costs on the interconnection, must be included. For local generation, account must be made for fuel, operating and maintenance charges, together with other such extraneous costs that may be incurred (eg housing).

All these costs are compared by present valuing at a test discount rate over the period of the study. It may be that, due to the incidence of capital expenditure in the above comparison, the method of meeting the new (or increased) load, which is the cheapest in the long term, may involve a relatively heavy cash flow in the early years. In this case, where limited borrowing facilities exist, it may be necessary to choose an arrangement which is not the cheapest in the long term but gives an acceptable short term cash flow.

COMPARISON OF FOSSIL FIRED GENERATION PLANT

In terms of electrical generation for small or relatively small isolated systems, there are two basic systems, ie fossil fuelled systems or natural power systems. It is not the purpose of this paper to consider the latter. The former category can be considered under three headings:

(i) Steam Turbine Generator
(ii) Gas Turbine Generator
(iii) Diesel Engine Generator

Steam Turbine Generator:

The use of steam turbines with direct boiler fired plant in the size of system under consideration (5-15 MW) is unusual unless there is a requirement for low grade heat using pass-out steam from the steam turbine. Without such a requirement the overall thermal efficiency will be of the order of 22%-25%. Furthermore, this efficiency will only be attainable at full load. The only practical advantage over the other systems is the ability of the boiler to be designed to consume various types of fuel. The capital cost of such plant is high in relation to the other systems due to the plant required.

Gas Turbine Generator:

Gas Turbine Plant is significantly lower in capital cost than similarly rated steam turbine plant and above 5 MW has considerable cost advantages over diesel engines. Their civil engineering and installation costs are lower and they do not require cooling water. The thermal efficiency at 20%-26% is, however, generally lower than comparable steam turbine plant and much lower than equivalent diesel plant.

The unit, being a rotating machine, does not require the degree of maintenance of a diesel engine (although it may ultimately require more skilled maintenance) and consequently should have a high availability. Owing to its high specific fuel consumption, as with the steam turbine, unless there is a waste heat requirement, the thermal efficiency of gas turbines in the range up to 15 MW will be of the order 18%-26%. There is, however, an application for gas turbines for standby or peaking applications.

Gas turbines in the present state of their design can only be operated on distillate oil fuels or gaseous fuels and, while designs exist for residual fuel burning, the capital and operating cost of such equipment normally mitigates against this use.

If, however, there is a waste heat application with a high heat to power ration, overall thermal efficiencies of up to 85% can be achieved with gas turbines.

Diesel Generator:

The diesel engine, which as an overall thermal efficiency of 40% (without any waste heat application) is therefore the obvious prime mover to consider for small isolated systems. It can operate on residual and distillate oil fuel. The maintenance requirement is more onerous than the equivalent steam or gas turbine but is less skilled than for such plant. As regards a waste heat application, it is not usually possible to obtain granter than 75% overall efficiency with a diesel. There is a set size limitation in that the maximum rating for diesels used in power generation is 15-20 MW, whereas gas and steam turbines are available in power ranges well beyond this.

The diesel engine is therefore the more usual unit for power generation in the capacities required for the Scottish Island systems and the alternative of Gas Turbines has not been employed except where gaseous fuels are available or where there is a requirement for standby plant.

TYPES OF DIESEL GENERATORS

Diesel engines are generally divided into three classifications by speed range, slow, say up to 300 rpm, medium (300-750 rpm) and fast (1000-1500 rpm). For base load generation it is more usual to stay in either the slow or medium speed ranges.

Both slow and medium speed engines can operate on residual fuels up to 70 cs (at 82.2°C) and in some instances above.

The low speed unit is more tolerant of impurities (sodium, vanadium etc) in the fuel, has fewer moving parts leading to less onerous maintenance requirements with greater availability and reliability. The maintenance charges on medium speed engines are some 50% higher than low speed units which also have a slightly higher efficiency, lower oil consumption and generally a lower sound emission, an important point in view of current legislation. Against this there is a high capital cost penalty on the low speed engine, particularly in the case of the actual generator and, further, its volume is considerably greater than a similarly rated medium speed unit, resulting in higher building costs. On the other hand there has been considerable development of the medium speed diesel over the past few years - speeds have increased as have mean effective pressures, and in general most manufacturers are obtaining approximately twice the power from current frame sizes than they were fifteen years ago.

The Hydro-Electric Board has consistently used medium speed engines in its diesel stations over the last twenty years. However, the maximum demand in a single station is at present of the order of 30 MW and it is not considered that for this demand slow speed crosshead units can be justified.

High speed diesel units are not in general acceptable for base load generation use and are only used for standby or peak lopping purposes. They are usually of lighter construction and are, usually, not able to burn residual fuels. Their low weight to power ratio does make such units suitable for temporary installations to cater in particular for overhead line or submarine cable failures.

It should be noted, however, that development work is proceeding, particularly in the 1000 rpm range at lower powers (0-2 MW) with residual fuel operation.

FUEL

Diesel engines can operate over a wide range of fuel oils from gas oil to residual fuel of 70 cs (at 82.2°C) and can be modified to burn gaseous fuel. In this case some pilot distillate fuel is required to maintain ignition, usually 5% and the engine will require to be down rated in the order of 5-10%.

Fuels for oil engines are classified in B.S.2869 and fall into two major classifications, distillate fuel and residual fuel. Distillate fuels have good handling properties except under exceptionally cold conditions - not generally experienced in the islands. The properties of residual fuels, however, are such that after the removal of distillate fuels and other refined products from the crude, the residue is viscous contaminated with trace elements, sand and other impurities. They are difficult to handle, require heated storage and treatment prior to combustion.

Nevertheless, the differential price between residual and distillate fuels has increased markedly over the past few years. The trend of this increase is represented in the following Figure which illustrates the increase over the past five years.

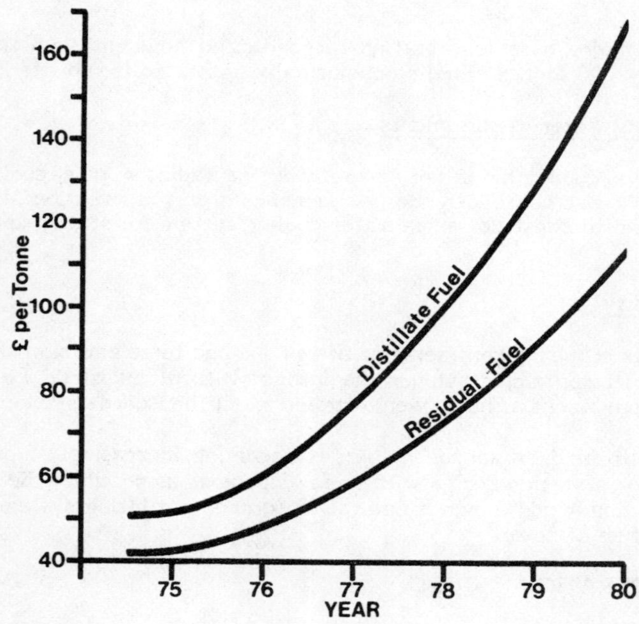

Fig 5 DIFFERENTIAL FUEL COSTS

In terms of island generation the following criteria have to be met for residual fuel operation. As the fuel requires to be delivered hot, the station must be situated within a reasonable distance of the port or harbour where it is off-loaded from tankers, usually 2 km, if unnecessary complications in the receiving pipeline are to be avoided. The pipeline itself must either be "pigged" after delivery or the line purged with a distillate fuel. The main fuel storage must be heated but such heating is usually obtained from waste heat sources. The fuel requires treatment prior to combustion and such treatment normally comprises heating to reduce its viscosity and specific gravity, followed by a centrifuging process, the treated fuel being stored in heated ready use tanks. Prior to combustion the fuel has to be heated to a temperature corresponding to a viscosity of 12-16 cs.

While the treatment process does incur extra costs for additional capital plant and maintenance, such costs are adequately compensated by the differential fuel costs.

SITING OF DIESEL STATIONS

The major factors in determining the siting of a diesel station are as follows:

(i) Access to Fuel - Proximity to Load Centre

These two criteria have to be considered in relation to each other. It is desirable that a diesel station should be situated relative to its load centre and at the same time the logistics of fuel supply require that, should the output of the station be in excess of 25 GWh annually, the station is sutated readily accessible to a deep water berth that can accept coastal tankers.

The Board's existing major base load stations are situated adjacent such facilities which allow cargoes of 1000-2000 tonnes. Fuel stock normally legislated for on site is six weeks.

(ii) Cooling Water Requirements

Diesel stations can be cooled by either evaporative or radiator type cooling towers or directly from a sea or fresh water loch source. Dependent on the situation of the station it is normally uneconomic to construct a sea water cooling system for station capacities below 2 MW.

(iii) Amenity

The two major aspects requiring consideration are smoke and fume emission and noise. With the use of residual fuel containing sulphur the Inspectorate of Industrial Pollution is now requiring relatively high stacks as have been installed at all the Board's stations.

Noise, particularly with medium speed engines, is becoming increasingly more difficult to deal with and calls for a station design with as few openings as possible, the elimination of all unnecessary doors and windows and a natural or forced ventilation system that in itself does not allow the egress of noise.

OPERATION AND OPERATIONAL COSTS

The Board's Diesel Stations are consolidated in a Diesel Group and are the responsibility of the Generation Engineer for Operation and Maintenance. Each individual station is in the charge of a Superintendent, responsible to the Generation Enginer, with two staff engineers (mechanical and electrical) and a foreman. The manual staff generally comprises ten operators, who operate the station on a continuous shift system (one operator and assistant operator per shift) eight mechanical fitters and mates, two electrical fitters and two labourers.

All maintenance and repair work is carried out by the station staff although contractors may be brought in from time to time to assist over peaks or on breakdown. The stations are essentially self-contained in terms of operation and maintenance.

Stations are situated within 2.5 km of harbour facilities and fuel supplies to all base load stations are delivered by sea.

The following block diagram indicates the average operational cost experienced in the Board's residual fuel diesel stations over a ten year period.

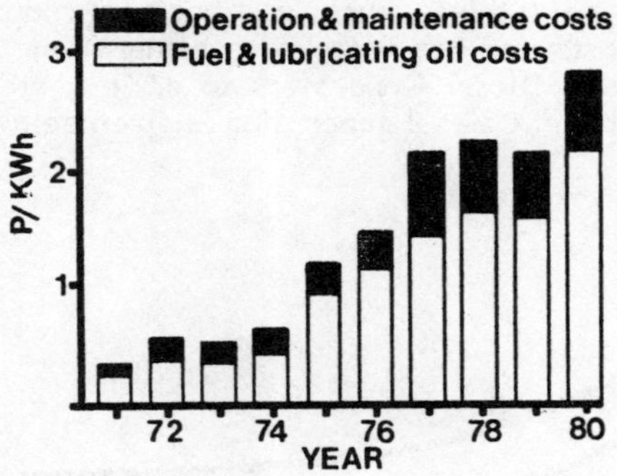

FIG 6 - OPERATIONAL COSTS

It can be seen that the fuel component represents about 70% of the total cost. Maintenance costs are the next highest single item, accounting for approximately 15% of the total cost. Lubricating oil, with medium speed engines is also significant being in the order of 5% while operating costs represent about 3% of the total cost.

FUTURE DEVELOPMENT

In the current energy situation it is accepted that the price of fuel oil will, in real terms, increase with a consequent increase in diesel produced power and therefore means will require to be devised of reducing the cost of electrical energy for communities that rely on diesel produced electrical power.

One method already practised by the Board is the burning of flare gas in a gas turbine at Flotta. In this installation gas is supplied on an interruptible basis and the plant cannot therefore be regarded as firm. Nevertheless, the gas availability has been reasonable over current financial year and an estimated 350 tonnes of residual fuel saved. This saving needs, of course, to be set against the capital charges incurred by the unit. Flare gas can contain higher fraction gases such as propane which, under certain circumstances, can condense out and cause damage to the turbine. Provision must be made to eliminate such a possibility.

A further energy saving scheme that can be adopted is the introduction of a total energy concept. Such schemes can be applied to diesel generation and can result in overall efficiencies in excess of 75%. However, such schemes on a small isolated system need to be carefully matched as the maximum demands of heat distribution and the electrical load system are unlikely to occur simultaneously and it would not be economic to use supplementary firing. There is little industrial or process heat requirement throughout the islands. Any district heating would be largely domestic and as the heating season is not particularly extensive its loading would have a low load factor. A variation would be to use the steam produced by waste heat in a small steam turbine generator and, while the overall efficiency is not so high, neverthless such electricity produced would be available throughout the year. Such a scheme has received consideration. A cost comparison using the waste heat from two 8 MW Diesel Generators to drive a steam turbine generator at various load factors against straight diesel generation is given below.

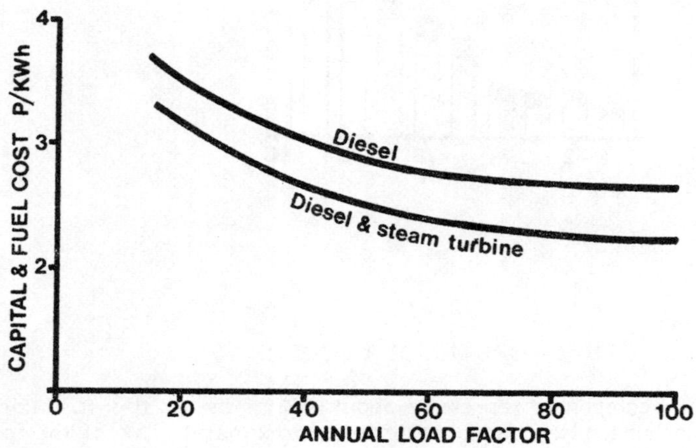

FIG 7

The annual costs include both fuel maintenance and capital costs for each system and even at low load factors savings can be made. It is estimated that at average load factors savings of 0.4p/kWh can be achieved. This is, however, assuming that the steam turbine generator is included in the "Firm Plant", allowing in the overall costing a capital cost reduction for equivalent diesel plant.

At present the major electrical demand on the Scottish Islands if supplied from fuel oil. With the present costs experienced for this form of generation and the knowledge that these costs may continue to escalate in real terms thee is no doubt that ways require to befound and are being investigated to temper these costs either by increasing the efficiency of the existing form of generation and by the introduction of other methods. At present on economic and practical considerations it is difficult to justify alternative forms of plant.

The author would like to acknowledge the assistance of his colleagues who assisted in the preparation of this paper.

Review of Developments in the Electric Vehicle Field

T. M. Baker

Energy Research Group, Open University, Walton Hall
Milton Keynes MK7 6AA, UK

ABSTRACT

A review is made of current developments in the electric vehicle field that may be of interest within a rural and island context. The review starts with a world overview then looks at UK development schemes. Those covered are "London goes electric", Cambridgeshire County Council electric mobile libraries, public passenger transport operations, cars and boats. The future prospects of electric vehicles are briefly touched on and details are given of the journal Electric Vehicle Developments as a source of further information.

KEYWORDS

Electric vehicle; battery; review; transport; rural; island.

INTRODUCTION

This paper makes a brief review of some of the current developments in the electric vehicle field. This has been done in the hope that it may provoke interest in electric vehicles within a rural and island context.

Probably the major drawback for electric vehicles are their limited range due to the high weight of batteries. For any vehicle use which can be achieved within their limited range there are usually substantial running cost advantages. It is likely that on many islands and within some isolated rural communities, the range limitations of electric vehicles would not pose a serious problem. For example, Neill (1980) has estimated that all of Hawaii's transport could be run with electric vehicles and that they would play an important role in helping Hawaii to become energy self-sufficient on renewable energy sources such as wind, geothermal, ocean thermal and hydroelectric.

As with the use of electric vehicles in Hawaii, their use in the Highlands and Islands could form a very useful load at a local level for small scale hydro, wind and tidal electricity generation. The advantage of battery charging as an electricity load is that it tends to be high when other loads are low (i.e. at night and in the summer) and it can, to a limited extent, take power when it is available rather than demanding it instantaneously like lights or television.

It is perhaps wind which is likely to be the most easy renewable source to tap within the near future. Lowe (1980) has estimated that the United States ERDA/NASA Mod 2 design wind machine could produce electricity at a cost of 0.96 to 0.54p/kWh for sites with mean wind speeds of 5 to 8 m/s (12 to 18 mph) respectively. This is with machines used in pairs. If there are five machines at one site the costs fall to 0.85 to 0.48p/kWh. The Mod 2 has a 91m diameter rotor and a 2.5 MW generator. As a comparison the cost of coal generated electricity is currently about 1.1 to 1.2p/kWh of electricity sent out.

REVIEW OF DEVELOPMENTS

In a report on the 14th Intersociety Energy Conversion Engineering Conference held in Boston, USA from the 5th to 10th August 1979, Öjefors (1979) reported a state of the art review of electric and hybrid vehicles. This review found that:

> The only long-term cost-effective major fleet operation of electric road vehicles in the world today is that of the milk delivery system in Britain.
>
> Japan is the only country where electric/i.c.e. hybrid vehicles are in daily commercial on-the-road use.
>
> In the USA the electrically powered vehicle has captured the interest and involvement of the major battery manufacturers and electrical components suppliers, but as yet only marginal interest has been shown by the major automobile and truck manufacturers.
>
> Major automobile manufacturers outside the USA have shown a more active interest in the development of the electrically propelled vehicle compared with their American counterparts.
>
> The only large-scale electric-vehicle fleet operating in the USA, that of the United States Postal Service, continues to operate satisfactorily in terms of performance on certain specific duty cycles.
>
> The light duty truck in fleet operations and specific duty cycles is the most common application of the electrically propelled road vehicle.
>
> Fleet operation appears to be the most appropriate means for the introduction of this new vehicle technology due to the need for central maintenance facilities.
>
> A necessary prerequisite to the successful introduction and effective utilisation of electric vehicles is the existence of a mature infrastructure which is composed of items such as electric power generation, transmission and distribution, charging facilities, maintenance facilities, replacement parts, vehicle maintenance and repair procedure documentation.
>
> Successful introduction of electric vehicles requires co-operation among vehicle manufacturers and suppliers of electricity.

Exxon Development Programme for an Electric Car

The major difference between electric and internal combustion engined (i.c.e.) vehicles is the density of the energy storage medium. In the case of the electric vehicle this leads to a critical trade-off between energy consumption and range. In the US, Exxon is working on the development of vehicles for the sub-compact market. They have found cost a greater limitation than weight, and as the range of a vehicle is increased the cost becomes very high. For all the battery systems they have examined, for maximum market penetration, the optimum range of an electric vehicle, which has equivalent performance to an i.c.e. vehicle, would be between 95 and 110 km (Pegg, 1980).

Government Support for Electric Vehicles

The largest programmes for the development of electric vehicles are in America and Japan where so far £80 million and £16 million respectively have been allocated by Central Government (Electric Vehicle Development Group, 1980). In the UK, government support has been more limited. Approximately £2 million has been given to Chloride Silent Power for the development of the sodium sulphur battery and a similar sum has been given to Lucas Batteries Ltd. for the development of electric vehicle components. The Department of Industry has also contributed £400,000 to the London Goes Electric programme (see below). In their report on electric vehicles the House of Lords Select Committee on Science and Technology says "Support should continue at its present level in real terms but does not have to be greatly increased. The government's objective should be to allow the EV to prove whether it can compete with the internal combustion engined vehicle on equal terms before oil supplies actually run down" (Stansell, 1980). Further government support for electric vehicles has been given with the abolition of Vehicle Excise Duty on electric vehicles. This will tip the balance of costs between electric and i.c.e. firmly in favour of electric vehicles in many instances.

DEVELOPMENT SCHEMES

London Goes Electric

Of all transport markets the local pick-up and delivery goods market is the most promising for the early introduction of battery vehicles. This is the field covered by the London Goes Electric project which is the UK's largest electric vehicle development and demonstration project. "The scheme has essentially two purposes:- to allow gaining of experience and development of reliable vehicles and then to demonstrate the performance of the vehicles" (Bayliss and Foulkes, 1979). Sixty-two vehicles have been leased or sold to operators at a price comparable to those of a diesel or petrol vehicle. Since the vehicles were produced in small runs they were more expensive to produce, however. This excess cost has been met by the Department of Industry. The programme is to run for three years and it is expected that the first year will have a significant element of debugging whilst the later years will demonstrate the reliability and operating economies of the vehicles.

Three manufacturers are involved in the programme and details of their vehicles are given in Table 1.

Preliminary results for Initial Services who have 3 Lucas-Bedford vans and all 12 Cromptons is that overall running costs are at least 30 per cent less than for their comparable diesel engined vans (Anon, 1979). The first results of the programme are being analysed and are due to be published in the autumn of 1980. One problem already encountered is with drivers. "A heavy footed driver could reduce the range by a factor of three and the reasons for this are being investigated" (Wicken, 1980).

TABLE 1 London Goes Electric Vehicle Specifications

Make	Crompton	Lucas-Bedford	Chloride-Dodge
Unladen weight (tonne)	2.56	2.5	4.5
Payload (tonne)	1.76	1.0	1.75
Range (km)	54	80	72-88
(mile)	(34)	(50)	(45-55)
Acceleration	0-14.5 mph 27 sec	0-30 mph 14 sec	0-30 mph 19 sec
Max.speed (m/s)	6.4	>22	18
(mph)	(14.5)	(>50)	(40)
Max.gradient fully laden	1 in 9	1 in 6	1 in 6
Further details:	R. A. J. More, Crompton Electricars Limited, Crown Avenue, Dukestown, Tredegar, Gwent, NP2 4ET	Lucas Batteries Ltd., Electric Vehicle Project, 17, Evelyn Road, Sparkhill, Birmingham B11 3JR 021-772 2277	W. N. Johnson, Chloride Technical Limited, Wynne Avenue, Swinton, Manchester M27 2HB 061-793 5000 x 490

Further work is being done on the development of the Lucas-Bedford van used in this programme. It will offer a breakthrough in terms of speed and range but production models are not expected to become available until late 1982 (Anon, 1979).

Cambridgeshire County Council Mobile Libraries

In August 1979 Cambridgeshire County Council introduced an electric vehicle into its fleet of mobile libraries. In May 1980 it confirmed that it was going ahead with plans to replace all 11 mobile libraries and four vans with electric vehicles. Monitoring of the electric vehicles is being undertaken by a committee which is chaired by the EVDG Executive member who is in charge of the London Goes Electric programme (Electric Vehicle Development Group, 1980). Comparative costs of i.c.e. and electric vehicles for this fleet are shown in Table 2.

For the mobiles there is a significant saving on the total cost (capital and running cost), but for the vans there is only a small saving. This is because the 'mobiles' are specialist vehicles which are individually constructed and so the electric versions are not greatly more expensive, whereas with the standard mass produced i.c.e. van the specially produced electric equivalent does cost significantly more to purchase. "Another point to be considered with the van is that the specification for the electric vehicle exceeds that of its i.c.e. counterpart" (Billingham, 1980b).

The main factor to emerge so far from the Cambridgeshire County Council programme "is the need for thorough staff training in the use of electric vehicles, not only the physical aspects of driving and maintenance but so that they understand why as well as how" (Allbut, 1980).

TABLE 2 Comparisons of Costs for Cambridgeshire County Council Mobile Libraries

Fuel type		i.c.		electric	
Vehicle type		Mobiles	Vans	Mobiles	Vans
Number of vehicles		11	4	11	4
Total annual	km	113,837	96,381	113,837	96,381
Average	km	10,349	24,095	10,349	24,095
Capital cost	£	220,000	28,000	264,000	48,000
Vehicle lives	years	10	6	20	20
Annual costs	£				
Depreciation		22,000	4,666	13,200	2,400
Licences		3,348	271	-	-
Fuel (1)		12,029	3,037	2,490	1,054
Tyres		835	445	835	445
Servicing & Repairs		10,360	2,270	4,950	2,400
Batteries (2)		-	-	5,775	2,100
Refurbishing (3)		-	-	5,500	2,000
Total annual cost	£	48,572	10,689	32,750	10,399
Cost per km	£	0.43	0.11	0.29	0.11

Notes:
(1) liquid fuel at £0.29/l (£1.30/gallon) and electricity at £0.0125/kWh;
(2) annual battery replacement cost based on an average battery life of 7 years;
(3) electric vehicles are expected to require refurbishing at 7 year intervals.

Source: Billingham (1980b).

Public Passenger Transport

There have been several electric bus demonstration projects. A major project was carried out by the South Yorkshire Passenger Transport Executive. The initial objectives were:

"1. To establish the present state of the art.
2. Determine the operational parameters and infrastructures necessary for extensive operations using battery vehicles.
3. Establish the power demand and consumption of these vehicles.
4. Following initial trials on two buses a further strategy of the programme was to rationalise the vehicle concepts in order to establish comparative reliability data" (Hellewell, 1978).

Four buses were involved, two 6.7m long 26-passenger buses built by Crompton/Willowbrook, the Chloride Silent Rider (a modified Seddon RU33), and the Lucas Electric Midi-Bus (based on a Seddon-Pennine IV chassis).

The major points which the drivers of the electric buses appreciated were:

"1. 'Automatic transmission' for driver convenience;
2. Smooth take off and precise handling, attributable to both the transmission and steering characteristics.
3. Improved braking characteristics, under the prevailing road conditions, i.e.

excellent" (Hellewell, 1978).

On the basis of the experience gained with the three types of bus, Hellewell (1978) estimated that the operating costs of diesel and comparable production electric buses would be about the same (see Table 3).

TABLE 3 Comparison of Operating Costs for Diesel and Electric Buses

		Expenditure per Passenger/km (p/km)	
Fuel type		diesel	electric
Operations/ Traffic	Salaries	21.7	21.7
	Battery depreciation	-	1.1
	Energy	2.2	1.7
	Sundry	4.3	4.3
		28.2	28.8
Servicing/ Maintenance	Vehicles	8.6	5.3
	Buildings/Plant	0.4	0.4
	Depot overheads	0.8	0.9
	Tyres	0.6	0.7
	Sundry	1.4	1.4
		11.6	8.7
Management	Insurance	0.3	0.3
	Administration	3.9	3.9
	Sundries,etc.	2.6	2.7
		6.8	6.9
Total		46.6	44.4

Source: Hellewell (1978).

More recently public transport operators in the UK have been talking rather than getting projects off the ground. They have also been looking at systems in operation in Europe. There is particular interest in dual mode (a combination of trolleybus and battery powered bus) and hybrid vehicles (which have batteries and an electric motor plus an i.c. engine either in parallel or series) (Munro, 1980).

Currently, Greater Manchester and South Yorkshire Passenger Transport Executives are carrying out test and development projects with three different types of electric buses. Also the City of Bournemouth has recently introduced a small electric bus service on one of its routes. This vehicle fits in well with the local environment, the running costs are low, and Bournemouth is now considering purchasing a larger vehicle (Electric Vehicle Development Group, 1980).

Cars

There are currently at least two electric cars being commercially developed in the UK. One is by a group of eight companies within the EVDG and the other is by Dragonfly Research Ltd.

EVDG electric car project. The criteria laid down for this project is for a maximum speed of 22 to 27 m/s (50 to 60 mph) and a range in the region of 160 km (100 miles). Due to the cost which would be incurred in producing a new body shell a Vauxhall Viva Estate has been chosen for the prototype. In the second stage of the project pre-production prototypes will be built with double skinned GRP body shells with high pressure, high density foam infilling. These will be purpose built vehicles using new technology for the suspension, body and chassis design, as well as new motors and control system specifically designed for them. The intention is that subsequent production will be in sufficient quantities to make it commercially viable and give a return on capital invested. It is also intended that the vehicle will be able to take on the new technology which is likely to come over the next ten years, (Billingham, 1980a).

The Dragonfly Nova. The aim of this project is to improve the 'image' of electric vehicles by developing a high performance hybrid electric car. In the first phase of the project a performance car was built. This was based on a track-proven Terrapin, a single seater club racer, which had a maximum speed of 38 m/s (86 mph). The next stage was to build a practical road vehicle using the data obtained. The resultant Dragonfly Nova is heavier, being a two seater, carrying lights, weather protection, carpets, windscreen wipers, and licence plates. The most significant weight increase comes from the need to carry an engine-generator set to provide mean runnin power while the batteries are limited to providing short duration peak power and acting as a store for regenerated power or surplus generated power. Running solely on its batteries the Nova has a range of about 32 km (20 miles) at an average speed of 22 m/s (50 mph). With the engine-generator in operation this is also the speed at which zero battery discharge occurs with a calculated fuel consumption of 3.7 l/100 km (76 mpg). In urban driving with an average speed of 13 m/s (30 mph) the fuel consumption is calculated to be about 2.8l/100 km (100 mpg). (Lee and Corbett, 1979).

Boats

Finally an area in which the weight of batteries is not a disadvantage is in boats because the majority of displacement boats carry ballast which can largely be replaced by the batteries. During the past four years Emsworth Shipyard in Hampshire has fitted six different craft with battery/electric propulsion. The two most successful applications lie in the field of workboats for small harbours and marinas and for cruising boats on inland waterways. A 7.9m (26ft) four-berth cruising boat used on waterways had a range at the limiting legal speed of 1.8 m/s (4 mph) of some 7 hours running time. It could be recharged overnight using normal 13A outlets. Gick (1980) "is firmly of the belief that battery/electric propulsion in boats has a great future."

FUTURE PROSPECTS

The first market which is likely to open up for electric vehicles is that of goods delivery vehicles. Progress in other markets is unlikely until the new high energy density batteries, such as Gulf and Western's Zinc-Chloride battery (Anon, 1980), are commercially available and/or the hybrid electric (such as the Dragonfly Nova) is developed successfully.

FURTHER INFORMATION

A good source of information on current developments is the EVDG's journal Electric Vehicle Developments which is generally accepted internationally as the leading magazine on electric vehicles and their potential use. It is published by the

Institution of Electrical Engineers, and provides a comprehensive worldwide coverage of technical developments in electric vehicles and related subjects, as well as giving information on operational experience.

ACKNOWLEDGEMENT

The author would like to thank the Electric Vehicle Development Group and its Director, Alan Aldous for help in preparing this paper. However any errors in the paper remain the responsibility of the author.

REFERENCES

Allbut, D. (1980). Minutes of the Annual General Meeting of the Electric Vehicle Development Group, 18th June, Electric Vehicle Development Group, London. p.3.

Anon (1979). Minister sees e.v. progress. Electric Vehicle Developments, 4, 7.

Anon (1980). Full speed ahead for the battery car. New Scientist, 86, 294.

Bayliss, D., and M. Foulkes (1979). Electric vehicles in a large city - future constraints and opportunities. In the proceedings of the Electric Vehicle Development Group Third International Conference, Resources for electric vehicles and their infrastructure, 20th-21st November, London. Peter Peregrinus Ltd., Stevenage. pp. 139-145.

Billingham, B. (1980a). Minutes of the Annual General Meeting of the Electric Vehicle Development Group, 18th June, Electric Vehicle Development Group, London. p. 6.

Billingham, B.M. (1980b). Fleet costings now favour electric. Electric Vehicle Developments, 6, 1, 4-5.

Electric Vehicle Development Group, (1980). Evidence for consideration by the House of Lords Select Committee on Science and Technology, Electric Vehicle Development Group, London.

Gick, P. (1980). Battery-electric propulsion in boats. Electric Vehicle Developments, 5, 9.

Hellewell, D.S. (1978). An operator's requirements and operational experience with battery electric buses and a consideration of future developments. In the proceedings of the Electric Vehicle Development Group Second International Conference, the economic use of electric road vehicles in a changing environment. 23rd-24th May, Sheffield. Peter Peregrinus Ltd., London. pp. 14-23.

Lee, G.A. and A.E. Corbett (1979). Development of a hybrid electric vehicle using high efficiency disc motors. In the proceedings of the Electric Vehicle Development Group Third International Conference, Resources for electric vehicles and their infrastructure, 20th-21st November, London. Peter Peregrinus Ltd., Stevenage. pp. 79 -87.

Lowe, R.J. (1980). The expected cost of electricity generated by the U.S. Mod 2 windmill. Energy Research Group, Open University, Milton Keynes. Presented at the British Wind Energy Association Wind Energy Workshop, April, Cranfield.

Munro, A.M. (1980). Minutes of the Annual General Meeting of the Electric Vehicle Development Group, 18th June, Electric Vehicle Development Group, London. p.2.

Neill, D.R. (1980). The Electric Vehicle Program in Hawaii. Presented at the EV Expo 80, 20th-22nd May, St. Louis, Missouri. EVC No. 8043. Electric Vehicle Council, Washington, D.C.

Öjefors, L. (1979). 14th IECE conference demonstrates worldwide interest in electric vehicles. Electric Vehicle Developments, 4, 16-18.

Pegg, R.E. (1980). Minutes of the Annual General Meeting of the Electric Vehicle Development Group, 18th June. Electric Vehicle Development Group, London. p.7.

Stansell, J. (1980). Lords plug a boost for electric vehicles. New Scientist, 87, 691.

Wicken, G.W. (1980). Minutes of the Annual General Meeting of the Electric Vehicle Development Group, 18th June, Electric Vehicle Development Group, London. p.3.

TOPIC C

Biomass and Ecosystems

The Ark Building Project on Prince Edward Island—Canada

Ken MacKay

The Ark Project, RR4, Souris, Prince Edward Island, Canada C0A 2BO

ABSTRACT

The Ark is an environmental designed house which includes facilities for horticulture, fish culture, energy conservation and the recycling of waste. The predominant energy input is solar radiation, although there is provision for wind power and wood burning support. Thus the project is an example of integrated holistic design for living and food production. The building has operated for one year, and the paper will present details.

KEYWORDS

Building; solar; horticulture; fish culture;

PAPER

Text of paper presented at the Conference not available for the published Proceedings.

Renewable Energy Supplies and Environmental Issues

Alan Mowle and Willie Williams

Nature Conservancy Council,
12 Hope Terrace, Edinburgh EH9 2AS, UK

1. ABSTRACT

All energy supplies have an effect on the environment, even if only to change the balance of longwave radiation to the atmosphere. This is part and parcel of the part played by energy supplies in maintaining economic activity in human communities. The important issues concern the nature of that impact on the environment and the balance between the needs of the economy and the needs of the environment. These are often in conflict. The environmental impact of renewable/conservation technologies would be fundamentally different from that of the energy supply scenario projected by present Department of Energy policies. The assessment of environmental impact should be an integral part of any policy process and an important determinant of planning decisions. The effects of renewable energy technologies tend to be immediate, limited in scale and non-persistent. The dominant energy technologies of the present-day tend to have long-term, subtle and persistent effects, exemplified by radionuclides, carbon dioxide and sulphur dioxide. This paper examines the nature of environmental impact assessment (EIA), reviews what is known of the impact of renewable technologies and comments on the implications for the planning process.

2. KEYWORDS

Renewable Energy, Environmental Impact, Energy Policy, Planning

3. EIA - the main issues

The term 'Environmental Impact Assessment' covers the assessment of the environmental implications of policies, programmes and development projects. The emphasis in the United Kingdom has been on projects rather than policies. Although guidelines have been published (DOE research report no.13, 1976) for the assessment of 'Major industrial installations', no criteria for appropriate scope of such studies has been agreed. At present there is no statutory requirement for EIA in the UK, in contrast to the USA, but this is the subject of a draft directive from the EEC (ENDS Report 53, 1980). The current draft would not require a mandatory EIA for individual renewable energy projects. A comprehensive EIA is a bulky document and will not be appropriate in all cases, but many existing reports on environmental impact limit their concerns to particular aspects of the problem, often those easier to handle or those generating most heat in the political process

EIA thus tends to be seen as part of the developer's public relations effort rather than as an integral part of the planning process. There are three categories of concern in EIA:-

3.1 Effects on human populations - The health and safety of workers and of human communities has recently emerged as a major component of EIA, encouraged by the implementation of the Health and Safety at Work Act, 1974 and the increased awareness of potential problems including cancers and other diseases, the possibility of explosions and other accidents. The important characteristic of these concerns is their power in the political process as a result of direct potential effects on the community.

3.2 Effects on ecological systems - These are often thought of as secondary to health and safety, but this relates to political perceptions rather than to functional reality. Ecological processes have no direct political voice, but these concerns derive from fears for the continued effective function of the biosphere as a framework for human activity; examples include the effects of CO_2 and SO_2 on the atmosphere. To that extent, concern about ecological processes is therefore fundamental to assessment of environmental impact. However, these processes are not well understood, and attempts to take account of them have resulted in some of the most bulky and indigestible of EIA reports.

3.3 Effects on aesthetic values - Concern with the visual and aural effects of an energy development may generate considerable heat in the political arena, and techniques have developed to take account of this. Some reports which pass as environmental impact statements consist of little more than an assessment of visual effects on the landscape. It should be obvious that these concerns are fundamentally different from those described in 3.1 and 3.2, and are functionally much less important. However, they are easier to assess and therefore tend to form a larger part of most studies.

There are two particular reasons why practitioners of renewable energy technologies should concern themselves with these issues. First, environmentally-conscious groups tend to be unselective in marshalling their arguments, and intensely conservative in their attitudes. Their opinion of novel technologies will be cautious at best, and may well be hostile if they have any doubts about the environmental impact. Second is a much more important point of principle. EIA tends to be project oriented for the historical reason that it has developed in response to particular problems with particular kinds of projects, and because a large part of the experience of such assessments comes from the USA, where formal development planning control is more rudimentary. As a result, there is no mechanism for comparison of the environmental impact of one technology against another, or of one energy supply strategy against another. As such assessments could provide powerful arguments in favour of renewable technologies ('soft energy paths' and so on), advocates of such technologies should give it more of their attention.

4. Environmental Effects of Renewable Energy Supplies - A Review

Before considering the role of environmental issues in the planning process, it is worthwhile to review the nature of the environmental effects of the main renewable energy sources. Limits of space and time dictate brevity on this occasion, so we have attempted no more than an overview of the present state of knowledge. We make a distinction between effects which are integral to the technology and those which depend on siting and design safeguards. This section is concerned with the former; consideration of the latter can only be done adequately on a project basis.

4.1 Hydroelectricity - Hydro is a long-established technology, although most public schemes are large-scale (more than 10^2 KW). Their effects are almost entirely limited to changes in flow regimes in the associated freshwater system and possible damage to salmon and trout populations where a dam is involved. Most of the problems are associated with dams, from the point of view of safety, ecology and aesthetics. The environmental effects of microhydro installations not involving a dam are thus correspondingly small. The use of pumped storage schemes to solve supply/demand matching problems for renewable supplies is much more likely to lead to problems due to fluctuating water levels.

4.2 Wave power - This technology is receiving the larger part of Government R & D spending on renewable supplies, perhaps because it is one potential source of supply which could generate large amounts of power (of the order of 1 GW) in a concentrated form similar to existing conventional power stations. An assessment of the environmental impact of siting energy convertors off the Outer Hebrides has been carried out by the NCC and indicates that direct effects on the environment would be very limited and unlikely to be harmful given careful design and execution. This would also be true of small-scale installations in other parts of the Highlands and Islands.

4.3 Tidal power - The largest operational scheme in Europe is at La Rance in France, and large sums have been spent on feasibility studies for a very large (ca.10 GW) scheme in the Severn Estuary. All this is based on a design concept of a funnel-shaped estuary with a large tidal range. There are no such sites in Northern Scotland, but a number of West coast sea lochs have significant tidal flows at their mouths. This rather different situation has not been closely researched, but there is no doubt that large energy fluxes are available. As a result, environmental impact is hard to assess, but we can say that a barrage would have serious effects on tidal movements and marine ecology, while tidal mills such as those proposed by Reading University would have a relatively small effect.

4.4 Wind power - This is another power source particularly suited to the Highland situation, but with a long and chequered history. Recent research has concentrated on developing high-speed machines suitable for electricity generation. With careful siting such machines would have a minimal ecological impact, but might pose problems of safety and aesthetics resulting from mechanical failure, ice formation and visual intrusion.

4.5 Solar power - The established technology involves roof-mounted panels producing warm water for heating purposes. While this has some application in parts of the Highlands and Islands, for example an installation in the youth camp building on Iona, its use will be limited by climatic factors. However, wider use of photovoltaics would follow the development of cheaper cells. At present, these are used only for specialised purposes, for example providing power for a telecommunications booster station at a remote location in Lochaber. Environmental impact of these sources is negligible.

4.6 Geothermal power - Usable geothermal sources have been found in Southern England, but no major sources have been identified in Scotland. Environmental impact is difficult to assess, but seems likely to be insignificant.

4.7 Biomass - This term covers a wide range of techniques, all of which involve exploitation of the products of photosynthesis, using vegetable matter in fermentation, pyrolysis or combustion to produce gases, oils or heat which may find a wide range of uses. Such sources will be of considerable use in particular local situations where supplies of raw material can be obtained at a competitive cost. None of the sources, which are the subject of a modest but wide-ranging programme of research funded by ETSU, are likely to provide a large part of national energy needs in future, largely because that would involve large areas of land needed for other purposes. Environmental impact is potentially significant at a local level, especially if marginal land which at present acts as an important reservoir for wildlife, both fauna and flora, is brought into intensive use. There are potential problems with the processing techniques used to convert the raw material into a usable fuel.

4.8 Peat - Peat is not really a renewable resource, unless one is working on a timescale of some 7000-8000 years. It is, however, usually included in reviews of novel energy sources for the Highlands, and has seen renewed interest in recent months. The potential resource in the Highlands and Islands is very large, perhaps of the order of 10^8 tons, and has been exploited for fuel on a small, local scale for hundreds of years. The environmental impact of such small-scale useage is usually unimportant, but operations on a larger scale could be much more significant. In Eire, commercial extraction of peat for energy purposes is well-established, and there are suggestions that such a technology should be introduced in a Highland context. This would involve identification and exploitation of the relatively limited areas of deep peat, unimpeded by rock outcrops and other obstacles. Such sites are limited in number, and many of these are designated Sites of Special Scientific Interest (SSSI). These are ecologically important as relatively stable and undisturbed wetland ecosystems of a type rarely found elsewhere in Europe. Unless these ecological considerations are an important element of the planned use of this resource, considerable opposition to large-scale commercial schemes is probable.

5. Environmental Issues and the Planning Process

The individual developer's direct contact with the planning process will be through the local authority planning control procedures. Most physical structures and extensions to existing structures, unless concerned with an agricultural or forestry enterprise, are subject to development control in the form of planning permission. We have deliberately avoided any attempt to describe this part of planning in detail, but we are trying to demonstrate that the part with which most of you will come into contact is only part of the whole, and that anybody interested in promoting novel technologies must have some understadning of all these parts if he or she is to succeed.

Local authorities (district councils except in Borders and Highland regions and the three Islands Councils, which are unitary authorities) deal with local planning (specific planning proposals put to them by developers) and formulate local plans based on the broad policy framework set out in Structure Plans. Structure Plans are produced by regional or islands councils and undergo various stages of public consultation before being approved by the Secretary of State for Scotland through the offices of the Scottish Development Department (SDD). Structure plans generally do not try to specify precisely what the likely future developments will be or where exactly they will take place, but establish a statement of preferences in terms of zones (for industry, for housing, for countryside conservation, etc.) and policy statements (for employment, for industry, for housing and other services, and for the environment) appropriate to the circumstances of the region. These are produced for five-year periods and are subject to public discussion and consultation with other statutory agencies in the drafting stages. If there are serious objections, the Secretary of State may call an examination in public (EIP) before a reporter. The reporter may recommend changes to the Structure Plan to the Secretary of State, who may in turn impose these changes on the Regional Council. In this way, objectors and Central Government may have a quite considerable effect on the direction of local authority planning policy. For example, Strathclyde region recently attempted to have Glen Fruin (between Loch Lomond and Gare Loch) designated as a potential site for petrochemical development, but this was firmly rejected by SDD. Again, Orkney Islands Council, despite their opposition to uranium mining, have been unable to specify their objection in the Structure Plan because of the Secretary of State's intervention following an EIP.

A planning application will be processed by the officials of the district council planning department, who consider the application in the light of the policies expressed in the Structure Plan and may consult other government agencies, including those with an interest in environmental protection. They are aided in this by a series of National Planning Guidelines (on Industrial Sites, Aggregates, Nature Conservation, Landscape, etc.) produced by SDD, which give criteria for referral to the Secretary of State. The decision on each application is made by the planning committee, formed of elected council members, who will accept, reject or defer the proposal on the basis of the advice of their officials and their own assessment of the desirability of the proposals. If the proposal is contrary to the Structure Plan or the planning guidelines, it may be 'called in' by the Secretary of State to be dealt with by SDD through additional consultation or may go to a public inquiry. Even when there is a public inquiry, the Secretary of State has the final decision, and he may or may not accept the Reporter's recommendations.

This is the sharp end of the planning process. Government may influence the nature of development in many other ways. Perhaps most obvious is the system of regional development aid exemplified by the Highlands and Islands Development Board. Such bodies try to influence the location of development by offering financial incentives for developments in disadvantaged areas. More subtle influences of government, exercised through control of the purse strings, may be identified in the nationalised industries. These include, of course, most of the energy industries. Large sums of money are involved here - perhaps the most striking recent example was the pressure placed on the CEGB to order the Drax 'B' power station, against its own wishes.

Energy supplies are fundamental to economic activity, and thus are subject to much more government intervention than many sectors of the economy. Government influences the development of energy technologies through allocation of research and development funds, and government assesses the prospects for matching supply and demand through the studies carried out by the Department of Energy. However, Government falls short of complete control of energy matters. There is no agency charged with a statutory responsibility to ensure efficient use of energy. Electricity boards are obliged to generate electricity, not energy, as efficiently as possible. They compete with the British Gas Corporation, the National Coal Board and the oil companies. An increase of turnover in each of these bodies is the conventional criterion of their success. Losers in these circumstances are energy conservation and other novel technologies, including renewable supplies, together with the needs of the environment.

The needs of the environment are particularly pertinent here. All public bodies are charged with the responsibility to:-

> "Have regard to the desirability of conserving the natural beauty and amenity of the countryside" (Section 66 of the Countryside (Scotland) Act 1967)

The precise interpretation of this obligation has never been laid down, and therefore it is difficult to assess how the responsibility has been discharged in practice. However, it is the only statement of intent directed at the integrity of ecological processes and of aesthetic consideration. Safety and pollution control affecting the human community are the subject of other acts; this stimulus is the driving force behind most legislation deriving from the European Commission. Within the planning process, the environmental input comes from a number of agencies, including the Nature Conservancy Council (NCC), the Countryside Commission for Scotland (CCS), Her Majesty's Industrial Pollution Inspectorate (HMIPI), the Health and Safety Executive (HSE), the River Purification Boards (RPBs) and the Department of Agriculture and Fisheries for Scotland (DAFS) among others, each of which can advise the local authority from its own particular viewpoint. Some local authorities and government agencies have an ecologist offering advice, but this is the exception rather than the rule. The problem is that the political and economic criteria have prominence over environmental criteria in the eyes of decisionmakers; the <u>direction</u> of development is either unplanned (deriving from 'market forces') or the subject of strategic policy studies which neglect environmental needs. Thus those bodies with particular concerns for the environment find themselves opposing specific developments for environmental reasons and making little progress because the development forms part of a wide-ranging policy drawn up without adequate reference to environmental needs. Similar circumstances affect advocates of renewable energy technologies. The successful substitution of renewable for non-renewable energy sources involves a different approach to energy use and therefore reform of the institutions as well as the technology. While the Department of Energy has to maintain an overview of energy matters, its influence over the allocation of public money to the various energy agencies is extremely limited. It therefore lacks executive powers over national energy policy.

Hence, a basic problem in both the environmental and energy use fields is the absence of government agencies whose powers cover the range of needs for the environment and energy respectively. In the USA, for example, the Environmental Protection Agency (EPA) acts as a focus for expertise and concern in environmental affairs; whereas our so-called Department of the Environment is little more than the combined functions of the previously-established Ministries of Transport, Housing and Local Government, and Public Buildings and Works.

In the absence of such appropriate agencies, it is difficult to see how the environmental impact assessment of alternative energy policies could ever be effectively achieved. EIA may therefore continue to be a rather restricted, project oriented technique which obscures rather than reveals the opportunities offered by renewable energy and energy conservation technologies.

6. Conclusions

This paper has been directed at practical technologists, in an attempt to review the inter-relationships between the renewable energy technologies, the environment and the planning system. It has also been directed at developers, in the hope that their attempts to get planning permission, or to persuade other public bodies to use their particular brand of expertise, will be helped by the perspectives we have offered. However, the paper is also aimed at planners, with the main observation that environmental criteria should be more systematically incorporated into the planning process, particularly at the strategic level, if an environmentally appropriate energy supply strategy is to evolve. At the root of the problem, however, energy and environment are two issues which urgently need coordinating agencies, with statutory functions defined by Parliament, which will act as a focus in each field for expertise and concern, and out of which appropriate policies may emerge.

NB The views expressed in this paper are the personal opinions of the authors and do not necessarily reflect the policies of the Nature Conservancy Council.

Wood Energy for Local Use

C. P. Mitchell

Forestry Department, Aberdeen University, St Machar Drive, Aberdeen AB9 2UU, UK

ABSTRACT

The use of wood as an energy source for local use in rural communities is examined. The most likely conversion route will be direct combustion. The sources of wood for fuel at present are scrub woodland, logging residues, early forest thinnings, sawmill and processing residues and in the future, energy plantations. Short rotation forest energy plantations can either use coppice or single-stemmed trees. For coppice, yields of 10 to 20 dry tonnes/ha/yr on 5 year rotations can be expected and for single-stemmed trees 6 to 12 dry tonnes/ha/yr on rotations of 12 to 20 years. Systems of growing trees for energy are discussed.

KEYWORDS

Wood fuel; charcoal; producer gas; direct combustion; logging residues; forest thinnings; mill residues, forest energy plantations.

INTRODUCTION

Wood energy to most of us is synonymous with firewood - for domestic consumption either in an open hearth or wood-burning stove. Wood however, can also be converted to more refined secondary, solid, gaseous or liquid, fuels. Generally the methods of conversion are either biochemical or thermochemical, the latter being the most important.

CONVERSION ROUTES

The thermochemical conversion routes use heat to degrade the complex chemical structure of wood into simpler and more convenient forms. Heating wood to a temperature of about 800°C in the absence of oxygen (pyrolysis) leads to two types of decomposition product (King, 1980):

gas - containing H_2, CO, CO_2, H_2O and small quantities of hydrocarbons and alcohols,

and solid char - containing refractory carbon and mineral matter.

These products form the basis of all the thermochemical processes. Of these, only two, production of charcoal and producer gas, are suited to small-scale local use. Other processes, such as production of synthesis gas, involve relatively large conversion plants and are more suited to industrialized areas.

Charcoal is produced by burning wood in a deficiency of air. Part of it burns to raise heat, which drives off the volatiles, leaving a char - charcoal. The major advantage of charcoal is that it is a high grade solid fuel. Although charcoal plays an important part in the fuel economies of many less developed countries, in energy conservation terms, it is extremely wasteful. Sixty to seventy per cent of the energy in wood is lost in charcoal production.

Producer gas is made by a process similar to that for charcoal, but with the significant difference that the emphasis is on collection of the gases, not the char. Producer gas has a low calorific value and is usually burnt on site, in boilers for steam raising, or in internal combustion engines to generate electricity. Wood gasifiers are being developed which are suitable for rural use, particularly for use in less developed countries (King, 1980). I see no reason why they cannot be used in the rural areas of developed countries.

Secondary fuels can also be obtained by the biochemical treatment of wood. Ethanol can be produced by such processes but the conversion plants and production costs are large. Such methods do not appear suited to rural communities.

Hence we are left with the possibilities of using wood as a solid fuel for direct combustion, as a secondary solid fuel - charcoal, or as a source of producer gas. Because direct combustion of wood is relatively straight-forward and the advantages of conversion to charcoal or producer gas are slight (except in specialized instances) then one might as well burn it.

Direct combustion of wood can be done for many reasons amongst which include:

- domestic and small-scale industrial space heating,
- crop drying,
- generating electricity.

Apart from the developments for use in less developed countries mentioned above the major developments in small-scale use of wood as an energy source is that in the Scandinavian countries. There forestry is an integral part of farming and the use of wood as a solid fuel is an every-day facet of life. The developments lie in the improving harvesting techniques, processing and handling systems (Hakkila, Leikola & Salakari, 1979; Ling, 1975; Mattson & Sundstedt, 1977).

Electricity generation using wood fuel is increasing in popularity. The city of Burlington, Vermont, USA public power utility uses whole-tree chips for fuel. It has proved so successful that a 50 MW wood-fired generating plant designed to burn 1500 tonnes/day of whole-tree chips has been commissioned (Cheremisinoff, 1980). Ireland and Sweden are also investigating wood-fired power plants. Such a system could well be tailored to suit island or rural communities - given an adequate supply of feedstock.

SOURCES OF WOOD FOR FUEL

There are four potential sources of wood for fuel at present. These are:

- scrub woodland,
- logging residues,
- early forest thinnings,
- sawmill and processing residues.

A fifth source may exist in the future - energy plantations.

Harvesting scrub woodland for fuel wood is not efficient because in scrub woodland wood production is not optimized and felling it for fuel generally rules out regeneration so making it non-renewable. However, it is a short-term energy source. Clearing scrub woodland for fuel and then replanting to optimize biomass production on a sustained yield/renewable basis is an obvious step forward.

The use of logging residues (most probably tops and branches although in some areas roots and stumps may be available) and early forest thinnings must necessarily be confined to areas where intensive forestry is practiced. Similarly, sawmill and processing residues will only be available where there are wood-processing plants. The use of hardwood logging residues to fuel wood-burning stoves and hearths in the south of England is already an existing market which is growing and causing the Forestry Commission concern (Crowther, 1980). Conifer logging residues are more suited to use in wood-burning stoves. There is very little difference between tree species as regards calorific value (18 to 20 GJ/dry tonne). The major factor differentiating tree species for combustion is moisture content.

Energy plantations, growing trees specifically for energy, can take two forms - coppice, which exploits the rapid early growth from stools with established root systems and plantations of trees with single-stemmed habit of growth. Theoretical studies of these two silvicultural systems of growing trees for energy have been supported by the Department of Energy. Aberdeen University (Mitchell, 1979) examined the potential of single-stemmed trees and the Forestry Commission (Pearce, 1980) looked at coppice. A four year experimental programme has just begun on the basis of the results of these studies. Yields of 10 - 20 dry tonnes/ha/year can be expected from hardwood coppice grown on rotations of five years and yields of 6 - 12 dry tonnes/ha/year from single-stemmed trees on rotations of 12 to 20 years. The five genera showing greatest potential for coppice production are Populus, Salix, Alnus, Nothofagus, and Eucalyptus. Species of Betula may also be suitable but their ability to coppice is not as well developed. One limitation of coppiced hardwoods is that they are very site-demanding which may limit their use in, say, the Highlands and Islands of Scotland. However, local clonal material of Alnus and Salix may have potential. The short rotation forest energy plantations in Sweden and Finland which are being established on peatlands (both cut-over and virgin) use clonal material of Salix species selected from natural populations. However, on the more exposed and less fertile sites longer rotations with single-stemmed trees would be more suitable. Here the full range of forest trees is open for selection - the species best suited to the site which gives the highest yield of biomass being selected.

Energy plantations may take a number of forms depending on circumstances. For instance, if a local community were to undertake such a cooperative venture then an area large enough to supply the energy needs of the community on a sustained basis will be required. If the population is relatively dense then perhaps district heating schemes using wood fuel may be used. For example, Kannus, a small Finnish county, has changed from total dependance on foreign oil to provide heat to a domestic heating system using locally obtained fuels. Of the total 250 TJ annual consumption of the area 30 per cent (75 TJ/ann) will be covered by the district heating system. The fuel consumption will be as follows:

wood chips	25 TJ/ann
sod peat	25 " "
sawmill residues	17 " "
heavy oil*	8 " "

* as a reserve for cold periods in winter (-25° to -35°C).

It has been calculated that, with this fuel composition, the price of the heat energy produced is one third of the price if heavy oil alone was used.

Management of large-scale energy plantations will involve mechanized harvesting and chipping of the trees in or near to the plantation. Automatic stoking of boilers from a hopper containing wood chips would add to the convenience. Such systems for domestic central heating systems exist and are being manufactured in Scandinavia.

Large-scale plantations (greater than 50 ha) may not always be suitable and here small areas of woods on farms could be managed to provide a continuous supply of fuel wood. A total production of 10-20 dry tonnes/ann would probably be sufficient for domestic space heating. Harvesting of trees can be manual and there are several small-scale chippers on the market that are powered by a tractor. In Sweden, for example, the annual heating requirements of a large farmhouse and outbuildings can be met from 50 tonnes of birch roundwood. The total requirement can be achieved by 4 man-days' work per year - the boiler being fed automatically.

A word of caution must be raised at this point. Studies in Sweden have revealed that storage of wood chips in an enclosed building can result in a build-up of fungal spores in the air with consequent risk to workers of 'farmers lung' (Thörnqvist & Lundstrom, 1979).

If no areas of land can be set aside specifically for energy plantations then existing or new shelterbelts can be designed so that shelter is maintained but portions of the belt are harvested for fuel.

CONCLUSION

Wood has a potential as an alternative energy source for rural and island communities. In the more remote areas, which can sustain tree growth, a farm may be self-sufficient for space heating using wood fuel. For larger communities wood fuel, produced on energy plantations, could be a useful supplement to the local energy economy.

REFERENCES

Cheremisinoff, N.P. (1980). Wood for Energy Production. Ann Arbor Science, Michigan.

Crowther, R.E. (1980). Lowland Broadleaved Woodlands and Silvicultural Research. In D.C. Malcolm (Ed.), Research Strategy for Silviculture. Proc. Discussion meeting, Institute of Foresters of Great Britain.

Hakkila, P., M. Leikola, and M. Salakari (1979). Production, harvesting and utilization of small-sized trees. Sarja B No. 46b.

King, G.H. (1980). Wood for Energy. Commonwealth Forestry Assoc. Lecture, April.

Ling, C.H. (1975). Firing equipment for wood residues. Institutionen for Skogsteknik Rapp. och Upps, 85.

Mattson, J.E., and E. Sundstedt (1977). Logging residue as fuelwood in forest farmers houses. Instituttionen for Skogsteknik Rapp och Upps, 121.

Mitchell, C.P. (1979). The Potential of Forest Biomass as a Source of Energy (with special reference to trees with normal single-stemmed habit of growth). Aberdeen University, Report to the Energy Technology Support Unit, Harwell.

Pearce, M.L. (1980). Coppiced Trees as energy Crops. Forestry Commission, Report to the Energy Technology Support Unit, Harwell.

Thornqvist, T., and H. Lundstrom (1979). Undersokning av diasporforekomsten i luften i samband med flislagring hos fliseldare i blekinge. Institutionen for Virkeslara. Uppsats, 93.

Fuels from Non-woody Plants

T. V. Callaghan, G. J. Lawson, R. Scott and H. A. Whittaker

Institute of Terrestrial Ecology, Merlewood Research Station
Grange-over-Sands, Cumbria LA11 6JU, UK

ABSTRACT

Despite the fact that the UK is densely populated, intensively farmed for food and sparsely forested, it is suggested that biofuels could contribute significantly to the nation's energy supply. Rural and island communities could approach self-sufficiency using fuels converted from agricultural crops and wastes, seaweeds and natural vegetation. The UK resource of natural vegetation is described in terms of widespread species, or opportunity energy crops, which could be harvested from the areas which they currently occupy, and highly productive, but less widespread, species which would be planted as dedicated energy crops. Possible management and harvesting regimes for these crops are considered and tentative costings derived.

KEYWORDS

Biofuels; energy-crop productivity; energy-crop distribution; energy-crop management; energy-crop harvesting; natural vegetation.

INTRODUCTION

While most people realise that our dwindling fossil carbon reserves are all products of past photosynthesis, it is less widely known that the amount of energy fixed annually by the world's vegetation is about ten times the world's annual use of energy. Furthermore, the total proven fossil fuel reserves are no more than the present standing biomass (mostly trees) on the earth's surface (Hall, 1979a).

International research on obtaining fuels from biomass (biofuels) has burgeoned during the past few years, with most effort concentrated on tree and agricultural crops and conducted in those countries with low population densities (eg. USA, Canada, Brazil, Sweden, Australia and New Zealand). The 'Solar Biomass Programme' of the UK Department of Energy (King, 1979) is an attempt to assess the potential contribution of biofuels in this small, cold, overcrowded, intensively farmed and sparsely forested island.

Photosynthesis is at best 5.5% efficient in capturing solar radiation (Hall, 1979b), meaning that maximum theoretical crop yields in this country are 134 t ha^{-1} yr^{-1} in the south-west of England and 89 t $ha^{-1}yr^{-1}$ in northern Scotland. In practice, because of climatic, soil and management constraints, agricultural yields do not exceed 24 t ha^{-1} yr^{-1} (1.3% efficiency). Although productivities for some conifers are reported up to 36 t ha^{-1} yr^{-1}, much of this production is shed annually and it is unlikely that the harvestable yield of a forest or coppice woodland will exceed 20 t ha^{-1} yr^{-1} (1.0% efficiency). It is interesting therefore that some small stands of natural vegetation have displayed productivities exceeding 30 t ha^{-1} yr^{-1} (maximum efficiency 2.2%) (Callaghan, Scott and Whittaker, 1980). These figures are more than could be expected for the same species in monocultures, but they do indicate that British natural vegetation, in ideal circumstances, can match the photosynthetic efficiencies of the most productive plants of the tropics.

Callaghan and others (1979) estimated the present production of plant dry matter in Great Britain to be around 181 x 10^6 t yr^{-1} (giving an average production of 8 t ha^{-1} yr^{-1}). This represents approximately 3260 x 10^{15} J (PJ) or 121.2 mtce (million tonnes coal equivalent), and is therefore 34% of the 1977 UK primary energy demand or 89% of the oil demand. They further estimate (accepting the large errors involved in the available plant production figures) that if dedicated energy crops were grown on all rural land, the total potential dry matter production of Great Britain would be around 385 x 10^6 t yr^{-1} (giving an average production of 17 t ha^{-1} yr^{-1}). This converts to 6928 PJ or 257.6 mtce and represents 72% of the 1977 UK energy demand or 188% of the oil demand.

Therefore, while Brazil could meet its entire energy demand from the plant production of only 4% of its land area (Hammond 1977), the UK cannot achieve self-sufficiency from energy crops without using the complete production of three times more rural land than actually exists. Even if the entire country were devoted to growing specialised and highly productive energy crops we could not hope for self-sufficiency. On the other hand, if even 1% of our current plant production were converted into energy, savings of more than £160 million in oil imports could be made. Furthermore, the above national conclusions need not be pessimistic in relation to the possibility of energy self-sufficiency in rural and island communities.

BIOFUEL SYSTEMS

The biofuel production system can be represented in terms of resource or energy pools, flows and gates.

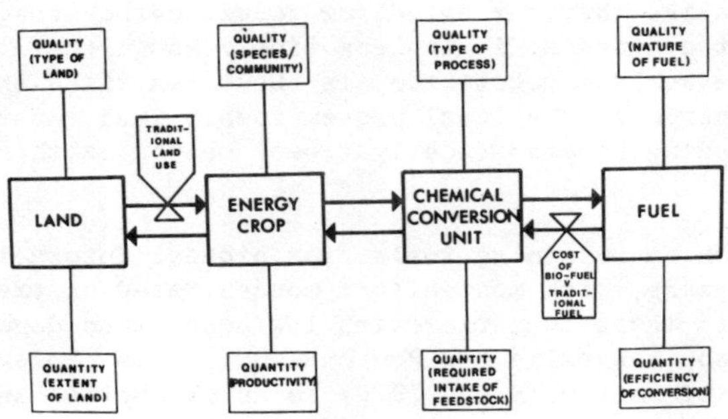

Fig. 1 Diagramatic representation of the biofuel production system

Land is the primary resource, and its area and fertility determine the yield and species composition of potential energy crops. Three types of chemical conversion process are envisaged: anaerobic digestion (Wheatley and Ader, 1979), thermal conversion (Ader, Bridgwater and Hatt, 1978) and alcoholic fermentation (Emery and Kent 1979). The selection of chemical conversion process depends both on the type of fuel required and on the quantity and quality of the available energy crop. The quantity and quality of the final fuel are largely determined by economic constraints related partly to costs involved in earlier stages of the production system and partly to the price of traditional fuels.

BIOFUELS FROM AGRICULTURE

Spedding, Bather and Shiels (1979) recently identified three possible agricultural fuel cropping systems

- Catch fuel crops grown in the time interval between the harvest of one crop and the planting of another, when the land would otherwise be bare.

- Break fuel crops grown in place of the less profitable crops necessary as breaks in cereal production.

- Plantation fuel crops grown on a larger scale, possibly separate from traditional farms, on land currently used for major agricultural crops.

Catch fuel crops must be capable of good germination and rapid growth in the few available months of late summer and autumn. Crops like kale, rape, fodder-raddish, vetch and even some weed species are most suitable, and the possible national yield, in energy terms is around 214 PJ for post harvest sowing or 311 PJ for aerial sowing (2.4% and 3.5% of UK primary energy demand). Energy costs would range from £1.2 to £4.4 GJ^{-1}.

Probable break fuel costs were established for forage rye (£1.4 - £1.9 GJ^{-1}), legumes (£2.1 - £2.3 GJ^{-1}) and ryegrass (£2.4 - £3.3 GJ^{-1}), and it is estimated that a gross energy equivalent of 7.6% of the UK primary energy demand could be achieved by replacing current crops of temporary grass and animal fodder by a fuel crop.

The most promising plantation fuel crop was identified as ryegrass, which could provide a biofuel feedstock equivalent to 41 - 48% of national primary energy demand if planted on all UK agricultural land.

BIOFUELS FROM AQUATIC PLANTS

Newman (1979) recently reviewed the potential of aquatic plants for energy cropping in the UK.

- Nationally, about 15,000 - 40,000 t dry weight of water-weeds grow annually in rivers, canals and larger drainage ditches. Of this, 10,000 - 20,000 t are currently cut, with an energy equivalent of 0.15 - 0.30 PJ yr^{-1}.

- Mass cultures of autotrophic algae are unlikely to be net producers of energy in this country, although mixotrophic growth on sewage or farm wastes is more promising.

- **Total UK seaweed resources represent 1.3 - 1.9 million tonnes ash-free dry weight with a potential annual yield of 0.3 - 0.5 Mt yr^{-1} or 6 - 9 PJ yr^{-1}. Trawl harvesting of *Laminaria* beds should have an energy input/energy output ratio of 1:10 or better. Since most UK seaweed grows off the coast and islands of NW Scotland, fuel from seaweed could be very significant in these regions.**

BIOFUELS FROM NATURAL VEGETATION

Natural vegetation is defined as that vegetation which is not grown for profit in agriculture or forestry. The category includes native plants and introduced aliens. Some species of natural vegetation are sufficiently abundant to provide *opportunity energy crops* without altering the existing land use; other species may be productive but are not sufficiently widespread to contribute significantly as opportunity energy crops. Such species would have to be planted and managed as *dedicated energy crops*.

The Resource of Natural Vegetation

TABLE 1 Maximum standing crops and extents of some species from unmanaged land

Common name	Latin name	maximum * standing crop t ha^{-1}	extent ≠ km^2
knotweed	*Reynoutria sachalinensis*	37.5	3.1
Japanese knotweed	*Reynoutria japonica*	25.3	-
reed	*Phragmites australis*	26.7	10.1
cord-grass	*Spartina anglica*	16.8	120
great hairy willow-herb	*Epilobium hirsutum*	15.3	6.0
yellow flag	*Iris pseudacorus*	14.8	0.1
rose-bay willow-herb	*Chamaenerion angustifolium*	10.8	138
creeping thistle	*Cirsium arvense*	10.8	334
policeman's helmet	*Impatiens glandulifera*	10.5	0.6
stinging nettle	*Urtica dioicia*	9.7	331
butterbur	*Petasites hybridus*	8.2	33.5
meadow-sweet	*Filipendula ulmaria*	7.1	110

≠ mainly Bunce (pers. comm.), calculated from the ITE Land Classification (Bunce 1980).

* Callaghan, Scott and Whittaker 1980.

85,610 km^2 in the UK, or almost 40% of all rural land, is classified as natural or semi-natural vegetation (Lawson, Callaghan and Scott, 1980). Because of minimal competition with other land uses, this area is expected to be the first to be made available for energy cropping. Initially this would involve opportunity cropping of unmanaged stands of vegetation, with heather moors (14,900 km^2 or 7.1% of rural

Britain) and bracken fells (3224 km^2 or 1.5% of rural Britain) providing the most likely candidates. The productivity of bracken may be up to 14 t ha^{-1} yr^{-1}, and heather, while its annual productivity is only around 2-3 t ha^{-1}, can produce a harvestable standing crop of around 15 t ha^{-1} at the end of a 15 year rotation period.

Several other species of unmanaged land could also provide opportunity or dedicated energy crops (Table 1), and productive use could be made of Britain's 1,200 km^2 of derelict urban and industrial land, 1800 km^2 of hedges and 2100 km^2 of road verges.

The Management of Natural Vegetation

Planting. Most potential dedicated energy crops of natural vegetation are perennial species. Propagation will involve transplanting of rhizome pieces during autumn and winter. Initially this may be done by hand but considerable mechanisation is possible using techniques developed in the Swedish Energy Reed Project (Björk and Granéli, 1978). Once a healthy stand is established repeated harvesting will be possible without incurring annual planting costs.

Harvesting strategy. The time and frequency of harvesting will be dictated by factors such as the possibility of regrowth within a growing season, the resillience of the vegetation to successive annual harvests and the need to replace, by fertiliser application, the plant nutrients removed by harvesting. The strategy selected will depend on the economic choice between maximising energy yield and minimising the energy input involved in management. Bracken and butterbur illustrate two possible strategies (Fig. 2).

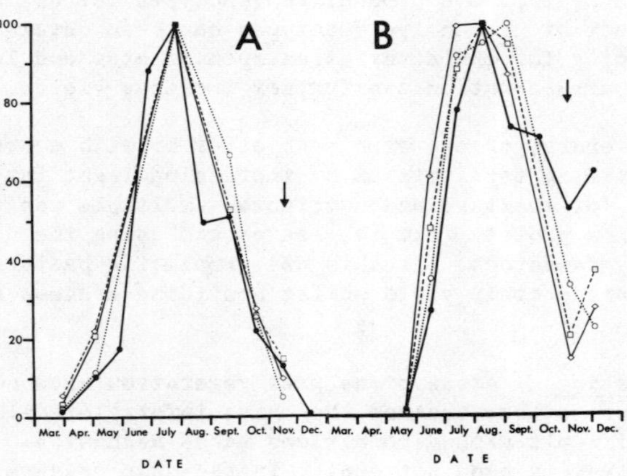

Note:

Thick solid line is standing crop.
Thin solid line is Phosphorus.
Broken line is Nitrogen.
Dotted line is Potassium.

Arrows denote the time of year by which all above ground tissues are dead.

Fig. 2. Seasonal changes in standing crop and mineral content in butterbur (A) and bracken (B)

- The total nutrient content in dying butterbur leaves decreases in parallel to the decrease in leaf dry weight, indicating that nutrients are not selectively moved, during autumn, to underground storage in rhizomes. Harvesting at peak biomass should therefore have little effect on regrowth in subsequent years, provided that soil nutrient levels are maintained by fertiliser applications.

- Bracken moves the bulk of its nutrient and carbohydrate reserves to underground rhizomes in late summer and autumn. Harvesting at peak standing crop would prevent this process and impede regrowth in subsequent years. Although the levels of important nutrients in bracken fronds are extremely low during autumn, the dry weight has decreased by only about 40% from its peak value. Thus, autumn harvesting is essential if bracken is to be managed as a permanent energy crop, and the yield penalty involved in foregoing the maximum standing crop is not excessive.

For some species, like bracken, reed and cord-grass, the envisaged chemical conversion route also has a bearing on the harvesting strategy. With harvesting early in the season the leaves are succulent and most amenable to anaerobic digestion, with later harvesting the leaves are dry and fibrous and are more suited to thermal conversion.

Fertilising. The harvesting of dedicated and opportunity energy crops will remove large quantities of nutrients which would otherwise have been returned to the soil through decomposition. Unlike agricultural harvesting, no information is available on the quantity and type of fertiliser application necessary to sustain the yield of repeatedly harvested natural vegetation. Callaghan, Scott and Whittaker (1980) assumed that fertiliser requirements would be equivalent to the weight of each element removed by harvesting, and found, particularly for end of season harvests, that required levels are generally below those used in intensive agriculture. The complete picture is complicated, however, and it will probably be necessary to assess the need for nutrient replacement by trial and error, as has indeed happened in agriculture.

Breeding. There is a wide range of variation in most natural vegetation species, and considerable increases over the average productivity observed in the wild can be expected merely by the selection of the more productive genotypes for use as dedicated energy crops. Once a stock of productive genotypes has been obtained it is expected that selective breeding for characteristics such as stem and leaf size, leaf density and longer leaf area duration can further increase yields.

Multiple land use. The mixing of energy crops, with each other or with conventional crops, offers the possibility of raising total yields by increasing light interception and decreasing competition for moisture and nutrients. Multiple use of a single crop is possible by expressing protein from the leaves and using the non-proteinaceous residues as an energy feedstock. In this way, replacing pasture with, say, Japanese knotweed would maintain protein yield whilst providing a cheap supply of biomass for fuel.

Technological feasibility of harvesting. Areas of natural vegetation have not been planted with conventional crops, either because they have infertile soils and unsuitable climates, or because difficult ground conditions makes mechanical harvesting uneconomic. These constraints need not apply, in the same measure, to energy crops of natural vegetation, which may be adapted to infertile soils and harsh climates and which can be harvested using a variety of unconventional techniques. Machinery is available to harvest the envisaged range of crops.

- Maize harvesters could be used for tall herbaceous crops, eg. Japanese knotweed, on flat, stone free land.

- Rotary slashers and flail mowers can cope with greater steepness and stonyness, with heavy duty models able to macerate woody stems up to 4" in diameter. Lighter models are very suitable for gorse and heather cutting.

- Powerful, low center-of-gravity, mini-tractors can operate cutting machinery on inclines of up to 40°, and thus cope with crops like bracken which often occur on steep slopes.

- **Low ground pressure harvesters, fitted with balloon tyres, are successfully used for reed collection, and could be employed for all soft ground crops.**

<u>Economic feasibility of harvesting</u>. It is very difficult to assess the economics of harvesting natural vegetation, but tentative calculations (Lawson, Callaghan and Scott 1980) indicate that costs of harvesting and transport to the farm gate, per barrel of oil equivalent, should be around £0.31 for heather, £1.12 for bracken and £7.45 for reed (£1.40, £5.09 and £34.01 respectively per tonne coal equivalent). None of these figures include the cost of converting harvested plants into useful energy, and the harvesting costs for heather and bracken are probably underestimates since they assume maximum yields and good ground conditions. Nevertheless, the figures do establish the comparative cheapness of energy from opportunity crops of natural vegetation, the more so if one discounts the current cost of burning heather moors (2.47 ha^{-1} - reducing the cost of energy from heather to £0.26/barrel of oil equivalent) and chemical erradication of bracken (£123 ha^{-1} - reducing the cost of energy from bracken to zero, and saving the farmer and the state an additional £85 ha^{-1}!).

THE POSSIBILITY OF ENERGY SELF SUFFICIENCY IN RURAL AREAS

The pattern of energy use in UK agriculture is reviewed in Leech (1975) and White (1977), and Slesser and Lewis (1979) analyse the economics and energetics of biofuel systems. We have concentrated on the biological aspects of biomass energy and have not had opportunity to make economic or energetic budgets for energy-crop systems.

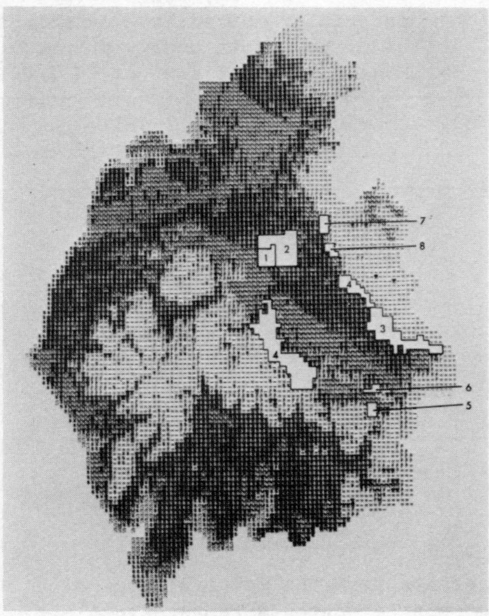

Darkest areas denote greatest plant production.
To feed alcoholic fermentation unit (assuming 50,000 dry t yr^{-1})
- area 1 = future energy crops in the lowlands
- area 1+2 = present vegetation in the lowlands
- area 3 = future energy crops in the uplands
- area 4 = present vegetation in the uplands

To feed anaerobic digestion unit (assuming 2,000 dry t yr^{-1})
- area 5 = future energy crops in the lowlands
- area 6 = present vegetation in the lowlands
- area 7 = future energy crops in the uplands
- area 8 = present vegetation in the uplands

Fig. 3. Possible catchments for two biofuel conversion units superimposed on map of present plant productivity in Cumbria

Nevertheless, we have identified, for the county of Cumbria, the likely areas of land necessary to meet the feedstock requirements of various chemical conversion units (Fig. 3). In the fertile lowlands of the Eden valley, for example, a catchment size of 15.04 km^2, growing conventional crops, could provide feedstock for a 5000m^3 digestor. This digestor would require 16,500 t yr^{-1} of plant dry

matter and could provide a nett energy output of 137000 GJ yr^{-1}, at a cost of £1.5 - 2.0 GJ^{-1} (Wheatley and Ader, 1979). This approximates to 91 GJ ha^{-1} nett energy output. A comparable figure using a productive energy crop (22 t ha^{-1}) would be 300 GJ ha^{-1}, which represents 46% efficiency of conversion of plant energy content to delivered gas.

An alternative to the above assessment of the potential energy production of land devoted entirely to energy production is to assume that any type of farm, whilst maintaining production of conventional agricultural products, can provide the equivalent of, say, 2 t ha^{-1} yr^{-1}, in crop or animal wastes or opportunity energy crops, as a feedstock for a small anaerobic digestor (Table 2). This analysis demonstrates that fuel and electricity self-sufficiency is quite feasible for most types of farm using only a nominal input of biomass. The cost of energy, at £4.9 GJ^{-1} is quite high compared with natural gas (£1.9 - £3.0 GJ^{-1}(1)) but is more reasonable when compared with liquid propane (£8.3 GJ^{-1}(2))or electricity (£6.2 - £14.9 GJ^{-1}(3)). Economics of scale operate with all digestors, perhaps up to a limit set by transport costs, and energy from 500 m^3 (£2.3 GJ^{-1}) or 5000 m^3 (£1.5 - 2.0 GJ^{-1}) models is much cheaper (Wheatley and Ader, 1979).

TABLE 2

% Farm self-sufficiency in fuel and electricity, assuming that any farm type can generate 2 t $ha^{-1}yr^{-1}$ in animal or plant material for use in a small anaerobic digestor

Type of farm	Fuel and electricity use $ha^{-1}yr^{-1}$* (GJ)	30m^3 digestor, 109.5 t yr^{-1} dry weight plant material required from catchment of 55 ha. Nett energy output = 800 GJ yr^{-1}, at cost of £4.9 GJ^{-1}	
		Energy use in catchment (GJ)	% Self-sufficiency
Specialist dairy	18.06	993	80.6
Mainly dairy	13.05	718	111.4
Cattle and sheep	7.65	421	190.0
Sheep	2.53	140	571.4
Pig and Poultry	33.20	1826	43.8
Cereal	10.80	594	134.7

* figures for average sized farms from Leach (1975)

ACKNOWLEDGEMENT

This paper has been made possible by research grants from the Solar Biological Programme of the UK Department of Energy. The paper's contents represent solely the views of the authors, which are not necessarily shared by the Department of Energy.

(1) Digest of Energy Statistics, 1980, for Aberdeen. Costs for 80 therms yr^{-1} and 800 therms yr^{-1}.
(2) Calor gas Retail Price List, 1980, not counting cost of cylinder.
(3) Digest of Energy Statistics, 1980, for Aberdeen. Costs for 75 kwh yr^{-1} and 750 kwh yr^{-1}.

REFERENCES

Ader, G., A. V. Bridgwater, and B. Hatt (1978). Conversion of biomass to fuel by thermal processes. Report to UK Dept. of Energy from Ader Associates, West Wickham, Kent.

Björk, S., and W. Graneli (1978). Energy reeds and the environment. Ambio, 7, 150-156.

Bunce, R. G. H. (1980). The application of computer aided techniques for landscape description in Britain. In S. W. Bie (Ed.) Quantitative Landscape Mapping in Western Europe: a Review of Experiences in the United Kingdom, Switzerland, West Germany, The Netherlands and Norway. Norwegian Computing Center Publ. 659.

Callaghan, T. V., A. Millar, D. Powell, G. J. Lawson (1979). Carbon as a renewable energy resource in the UK - conceptual approach. Report to UK Dept. of Energy from Institute of Terrestrial Ecology, Cambridge.

Callaghan, T. V., R. Scott, and H. A. Whittaker (1980). The yield, development and chemical composition of some fast growing indigenous and naturalised British plant species in relation to management as energy crops. Report to UK Department of Energy from Institute of Terrestrial Ecology, Cambridge.

Emery , A. N. and C. A. Kent (1979). Fermentation of biological material to alcohol. Report to UK Dept. of Energy from Birmingham University, Biochemistry Department.

Hammond, A. L. (1977). Alcohol: a Brazilian answer to the energy crisis. Science, 195, 564-566.

Hall, D. O. (1979a). World biomass: an overview. In Biomass for Energy, conference C20 of International Solar Energy Society (UK Section), pp. 1-14.

Hall, D. O. (1979b). Solar energy use through biology - past, present and future. Solar Energy, 22, 307-328.

King, G. H. (1979). UK Department of Energy Solar Biological Programme - Biofuels. In: Biomass for Energy,conference C20 of International Solar Energy Society (UK Section), pp. 15-25.

Lawson, G. J., T. V. Callaghan, and R. Scott (1980). Natural vegetation as a renewable energy resource in the UK. Report to UK Dept. of Energy from Institute of Terrestrial Ecology, Cambridge.

Leach, G. (1975). Energy and Food Production. IPC Press, Guildford, Surrey.

Newman, D. A. (1979). Aquatic plants as an energy crop in the UK. Report to the UK Dept. of Energy from University of Sheffield, Botany Department.

Slesser, M. and C. Lewis (1979). Biological Energy Resources. E & F. N. Spon, London.

Spedding, C. R. W., D. M. Bather, and L. A. Shiels (1979). An assessment of the potential of UK agriculture for producting plant material for use as an energy source. Report to UK Dept. of Energy from Reading University, Dept. of Agriculture and Horticulture.

Wheatley, B. I. and G. Ader (1979). Conversion of biomass to fuels by anaerobic digestion. Report to UK Dept. of Energy from Ader Associates, West Wickham, Kent.

White, D. J. (1977). Prospects for greater efficiency in the use of different energy sources. Phil. Trans. Roy. Soc. Lond. B. 281, 261-275.

Wood as a Source of Energy for Rural and Island Communities

A. I. Fraser

International Forest Science Consultancy,
Penicuik, Midlothian, Scotland, UK

ABSTRACT

The properties of wood as a fuel for domestic or industrial purposes are examined and a number of advantages identified which have particular relevance to island and rural communities. Wood is shown to be one of the cheapest sources of domestic fuel in Scotland and sufficient low grade wood is available at present to supply about 2 per cent of the rural population of Scotland. This proportion could be increased to 12 per cent within a decade if much of the 100 000 ha of hardwood scrub were converted to productive fast growing crops of Birch, Alder or Sycamore. The currently available technology for converting the energy stored in wood to more convenient forms is reviewed and some capital cost comparisons made of alternative systems.

KEYWORDS

Energy; wood; rural communities; fuelwood plantations; domestic energy.

INTRODUCTION

Wood has long been used by rural communities as a major source of energy and in many countries still represents the most readily available and important form of domestic energy.

In the U.K. wood has declined in importance, partly because of abundant supplies of coal and more recently oil, and partly because of destruction of forests to make way for sheep which were evidently a better economic proposition, especially when Britain had colonies to supply cheap wood.

Fossil fuels seem set to rise steadily in price in real terms because of the great increase in investment needed to extract each additional unit, with the result that interest is again focussed on potentially renewable forms of energy. Wood is a renewable form of energy which lends itself well to being used by rural or island communities where population densities are low and land may be available for tree planting.

The principle attributes of wood as a source of energy are:

a) It is renewable provided supplies are obtained from properly managed fuelwood plantations.
b) It can be easily stored.
c) It contains no polluting impurities such as sulphur.
d) It can be readily converted to other more convenient forms such as heat, electricity or gas, and technology may soon be available for converting it to liquid fuel.
e) It can be produced cheaply in small quantities to meet individual requirements.
f) The technology for harvesting and converting wood is simple, low cost and well proven.
g) Fuelwood plantations can be aesthetically pleasing and can bring other benefits such as shelter, wildlife.
h) The energy can be extracted from the wood after it has been used for other purposes or from residues produced during conversion processes (e.g. sawmill offcuts).
i) The ash residue resulting from burning wood is a very low proportion of the original fuel weight (1.5 to 3.0 per cent) and can be recycled as a fertiliser.

The average proportion of household expenditure going on heat, light and power in Scotland is about 6 per cent and this is likely to be higher for rural and island communities which rely on fossil fuels because of the high cost of distribution.

The big argument in favour of wood for isolated communities is that it can be produced locally and helps to increase the degree of self-sufficiency as well as reducing the disruption caused when supplies are delayed by adverse weather, industrial inaction, etc.

THE SUPPLY OF WOOD FOR ENERGY

In the short term of 5 to 10 years supplies of wood energy are determined by the amount of woodland in existence at the present time and the growing stock that it carries. In the longer term, that is ten years from now, the supply of fuelwood will be determined by action taken now to establish fuelwood plantations.

In 1965 the Forestry Commission carried out a Census of all woodlands and produced an estimate of the area of hardwood scrub in Scotland of 100 000 ha. There is of course an additional 37 000 ha of commercial hardwood forest and 700 000 ha of predominantly coniferous forest which could supply fuelwood but are likely to have more valuable uses. The Census gives an estimate of the growing stock of the utilisable scrub of 400 000 m^3 or about 100 m^3/ha. The unutilisable scrub volume is not estimated, but is likely to range between 5 and 50 m^3/ha.

A stocking of 15 m^3/ha might therefore be taken as a reasonable average figure giving a total growing stock of around 1.5 million m^3, or about 1 million tonnes of dry wood. This has an equivalent energy content to about half a million tonnes of coal.

This scrub woodland is almost certainly stagnant, with mortality and other losses equalling or exceeding growth so that harvesting is not possible without a reduction in growing stock. If it were harvested over 10 years, it could supply the equivalent of about 50 000 tonnes of coal annually, or about 5 per cent of the annual domestic consumption of solid fuel, worth about £3 million annually to the consumer.

On a national scale the amount of domestic energy that could be supplied from the existing wood resources is relatively small, being perhaps 10 000 families or about 1 per cent of the population. At a local level it could be a very important source of energy for isolated communities who have areas of scrub woodland nearby and must incur a high cost in bringing in alternative fuels.

In the longer term the possibilities of using wood for domestic energy for rural households and small institutions such as schools, public buildings and hospitals is quite promising. Most of the land currently carrying unusable scrub is capable of producing substantially more wood if it were converted to intensively managed coppices of improved strains of relatively fast growing hardwoods such as Alder and Birch.

Both these species are capable of producing about 5 to 10 tonnes of dry wood per hectare per annum so that over half a million tonnes of dry wood could be grown annually within the existing wooded area - a five-fold increase over the possibilities with the present resource. The rural population of Scotland is about 1.5 million so that the 50 000 families which could be supplied with wood represent about 12 to 15 per cent of those that might be in a position to benefit - a significant proportion.

In addition to the land currently carrying scrub woodland there are substantial areas which might be planted up with hardwood coppice to supply fuel for local communities. One of the most promising long-term possibilities for establishing fuelwood as a renewable resource is to plant trees in worked-out peat bogs. This is a technique which has been applied in Finland and the USSR and is currently being investigated in Ireland. Drainage and cultivation of the mineral substrate are likely to be necessary and the annual dry matter yield is much lower than that obtained by removing a layer of peat over the same area but these disadvantages are at least partly compensated for by the fact that the wood supply is renewable, provided the coppice stools are replaced from time to time.

Other types of land which lend themselves to fuelwood crops for local use are hedgerows and shelterbelts, river banks, roadside verges and small parcels of land which are not needed for other purposes.

THE PROPERTIES OF WOOD IN RELATION TO ITS USE AS A SOURCE OF ENERGY

Wood like most fuels is a variable commodity but some properties are sufficiently consistent for typical values to be quoted and used in design work. Wood cellulose has a heat value of 4 610 kcal/kg (19.3 MJ/kg) and most woods have between 370 and 450 kg dry weight/m^3. Standing wood has a moisture content of about 50 per cent which may be reduced by air drying to about 15 to 20 per cent, and the heat value of woodfuel depends on its moisture content. There are not great differences between species, with the exception of highly resinous conifers which can have heat values about 10 per cent above those of other species. Pine resin has a heat value of about 9 388 kcal/kg (39.3 MJ/kg) according to Powers (1978).

The heat value of air dry wood (15 per cent m.c.) is about 3 820 kcal/kg (16 MJ/kg) and for fresh wood (30 per cent m.c.) is about 3 350 kcal/kg (14 MJ/kg). For practical purposes a figure between the last two would represent a reasonable value for fuelwood in most rural and island communities in the U.K.

Table 1 compares the heat values of wood with other common domestic fuels.

Table 1 Comparative Thermal and Energy Properties of Common and Domestic Fuels

	kcal/kg	MJ/kg
Natural gas	11 310	47
Heating oil	9 565 - 10 520	40 - 44
Coal	6 450 - 8 600	27 - 36
Wood charcoal	6 700	28
Wood (20 per cent m.c.)	3 590 - 4 610	15 - 19
Peat	1 670 - 2 870	7 - 12

The table shows that wood has about one-third the heat value of natural gas and fuel oil, about one-half that of coal and about twice that of peat.

The economics of using wood as fuel are mainly dependent on the cost of growing and harvesting it and converting it to a convenient form for use in whatever conversion system is being operated.

The growing cost is mainly determined by the site conditions and the tree species chosen. Fertile lowland sites which allow high growth rates and species which have relatively fast growth rates in their early years will give the lowest cost of fuelwood. Birch, Alder and Sycamore are examples of relatively fast growing hardwood species. The harvesting cost depends on a variety of factors, of which accessibility, the number of trees per unit area and average tree size are the most important. Limbwood from large open grown trees being of irregular size and shape is costly to harvest. The most economic method of producing fuelwood is likely to be in dense plantations where trees will tend to be fairly uniform and can be harvested when the trees are at the most convenient size for the subsequent conversion technology.

The preparation of wood as a fuel must be well organised if costs are to be kept down and it is preferable to harvest the wood well ahead of the time when it will be used as fuel to allow plenty of time for drying by sun and wind. At the present time the woodfuel supply in Britain is disorganised with the result that the consumer who buys wood has no idea about the quality or heating value of his fuel. There is no reason why wood should not be converted from the tree length into billets or lumps to suit any conversion technology and dried to a standard moisture content so that it can be directly compared with other fuels.

Wood has the advantage over all other domestic fuels, except peat, that it can be harvested and prepared by an individual family or community to their own requirements. If cultivated locally, trees can be grown on short rotations of 7 to 10 years to 10 to 20 cm diameter and 5 to 10 m tall, which thus require a minimum of additional conversion apart from cross-cutting to render them suitable for burning.

Felling small sized trees is straightforward with a chainsaw and farm tractors can be used to drag bundles of small trees to a convenient landing. Cross-cutting can be done with a tractor mounted circular saw and no further splitting is required if small trees are used.

For larger consumers such as an institution, wood chips may be preferable and there are now available a number of mobile chipping machines which can reduce the tree stems to a uniform size suitable for a self-stoking boiler. Several models of mobile chipper are available which are claimed to be able to chip about 5 tonnes per hour of wood so that one machine can produce sufficient chips to provide a continuous output equivalent to about 4 MW.

WOOD ENERGY CONVERSION TECHNOLOGY

Until recently wood has been used as a fuel mainly by wood industries who have ready access to an abundant supply of low cost fuel. In the U.S. it has been estimated that nearly 50 per cent of the energy consumed by wood processing industries is derived from wood residues.

Only in very recent years has wood again gained popularity in Britain as a domestic fuel but few industries or small establishments give any consideration to using wood as a fuel.

For domestic users there is a wide choice of commercially available appliances which can burn wood but there is rarely information on their efficiency and each model has different requirements for fuel in the way of preferred length and diameter. The choices broadly range between stoves which are individual room heaters relying mainly on radiation and convection, through boilers with water jackets to provide heat for small bore water or hot air central heating systems, to ranges of various types which can also be used for cooking.

A single stove, with an output of 4 to 6 kWh capable of heating a large room will use about 3 to 4 tonnes of wood annually, while a boiler or range with an output between 25 and 30 kWh will use 10 to 15 tonnes of wood annually under upland conditions in Britain so that maintaining and storing an adequate supply of wood may cause problems for some but in rural areas should not create difficulties.

For industries or local authority establishments such as schools and hospitals there are a number of larger boilers which can provide either process or space heat or steam. Most are normally supplied with self-stoking equipment which means that the wood must be converted to chips and stored in a hopper which increases capital costs, but the degree of automation so obtained greatly reduces the cost of tending the equipment.

In the past few years the fluidised bed technology has been developed to a point where a number of models are commercially available which can burn wood and they have the advantage that they can handle green chips more efficiently than the traditional boilers.

The more conventional types of boiler generally employ a spreader stoker and fairly large lumps of wood in the range 3 to 5 cm are preferred compared with the fluidised bed system which requires small chips around 1 cm.

A third system available to the larger establishment is a wood gasification plant which can use wood or other waste organic matter to produce a low calorific value gas suitable for driving a diesel generator. The conversion efficiency for the gas generator is claimed to be about 75 per cent and overall efficiency is 25 to 30 per cent so that a plant consuming 20 tonnes per day or dry wood will give about 1 MW of electricity. Such a plant has been installed on an island in the South Pacific and may offer some advantages.

Other technologies are currently being developed to produce a high calorific value gas which could be comparable with Natural ("North Sea") gas and a range of liquid fuels such as methanol or a heavier fuel oil. Plants are in operation using these processes in the U.S. and Brasil but it is likely to be a few years before the technology has been refined enough to make the economics attractive.

The choice of fuel and the choice of system ultimately depends on the economics and some comparative cost figures for different fuels and conversion plants are of interest.

Current delivered prices for domestic fuels in Central Scotland are given in Table 2.

Table 2 Comparative Prices for Major Domestic Fuels in Central Scotland (1980)

Fuel	Cost per tonne (£)	Cost per equivalent kWh (£)
Electricity	-	0.0337
Fuel oil	160	0.0136
Coal	70	0.0079
Wood	30	0.0063
Natural gas	(16.50 per 100 Therms)	0.0056

In most rural and island communities gas is unlikely to be available and coal is likely to have an additional transport surcharge so that wood is clearly a fuel that bears very close study for these communities. There are currently large volumes of first thinnings all over Scotland which could well provide an abundant source of low cost fuel for many rural dwellers.

The cost of £30 per tonne used for wood in the comparison in Table 2 is based on delivered prices for pulpwood or chipwood, with some allowance to cover the cost of additional preparation.

Current costs of various wood burning installations are given in Table 3.

Table 3 Capital Costs of a Range of Wood Fired Appliances (1980)

	Output kWh	Cost £	Capital cost per kWh (£)
Stoves	2 - 8	200 - 500	60 - 100
Domestic wood-fired central heating system complete	25 - 35	2 000 - 3 500	80 - 100
Wood gasification plant	1 000	55 000	55
Industrial wood-fired boiler	1 700	26 000	15
Industrial boiler plus steam electricity generator	400	300 000	750
Wood gas plant plus	100	40 000	400
diesel generator	1 000	300 000	300

The technology is available for converting the energy stored in wood to other more convenient forms. Wood therefore must be given very serious consideration by rural and island communities as a means of improving living standards while maintaining a degree of independence from the problems created by delays in supplies and rising distribution costs.

REFERENCE

Powers, T.J. Preliminary environmental assessment of biomass conversion to synthetic fuels. EPA-600/7-78-204 US Environmental Protection Agency: Cincinnati, 1978.

TOPIC D

Wind Energy

Wind Energy Research Experience of the Reading University Energy Group

N. H. Lipman, P. D. Dunn, P. Musgrove, E. Bossanyi, G. Whittle, G. Slack, E. Preston and T. Eisa

The Energy Group, Department of Engineering, University of Reading, UK

I am proposing to give a general talk on wind energy. But I will be using results obtained by the Reading Energy Group, by way of illustration. At Reading we have one of the largest energy research teams in the U.K. The group is based in the Department of Engineering and is thirty persons strong. We collaborate with some ten researchers from other departments of Reading University. In addition eight members of the ITDG (Intermediate Technology Development Group) are based at our "Shinfield Road" outdoor test site. ITDG is an independant organisation but in practice they and we work very closely together on energy R. & D. Wind energy research represents about $\frac{1}{3}$ of our total activity. The names of the people involved at Reading are listed in figure 1.

I will assume that many members of this audience have not been privy to the detailed dialogue on Wind Energy that has been going on for the past three to four years, and I therefore intend to start from the beginning. For the proponents of wind energy in the U.K. it has been a long and hard struggle trying to convince the "Authorities" that wind is an alternative worth taking seriously. It has been equally difficult obtaining the small sums of money necessary to carry out our researches. But in the past year the situation, with regard to the D.En. and the electricity authorities has greatly improved. I believe that our Reading Group (plus a number of groups at other universities) has played a significant role in bringing about this change of heart. This is encouraging to those who feel that universities have a useful role to play in the nations economic affairs!

It is fair to say that Wind Energy is now seen as the "front runner" in the alternative energy stakes in a number of countries. Countries with particularly advanced wind R. & D. programmes include the U.S.A., Denmark, Sweden and Germany. In each of these, megawatt size windmills are being tested or will be shortly. Each has a number of industrial firms involved in the construction and marketing of both large and small windmills. In figure 2, I show one of the large U.S. windmills - the 2 MW G.E.C. machine at Boone in North Carolina. In the U.K. our programme is at least two to three years behind. We do not have a single windmill available for R. & D. or for deployment. I see a serious danger that the U.K. will be a non-starter in the race towards windmill commercialisation. However, there are some more encouraging signs, some of which I list below:

1) The recent CEGB announcement that it hopes to install a 100 KW windmill by next year and a megawatt machine by 1985 on an inland site.

2) The North of Scotland Hydro Boards stated intention to install a 20 KW windmill on Orkney as the start of a hoped for larger programme.

3) The D.En. funded design study of a 3.7 MW machine (60 metre diameter) that will, hopefully, be constructed on a windy hill top site in the South of Scotland. The word "hopefully" means that a decision to go ahead is still awaited! I hope that we will not have to wait too much longer, as the firms concerned - Taylor Woodrow and B.A.C. have already had a very long and slow ride with D.En.

4) A design study for a 25 metre version of the Musgrove, vertical axis windmill has recently been funded by D.En. McAlpine are leading this industrial consortium.

5) A first study of North Sea clusters of windmills has been carried out by Taylor Woodrow and its collaborators. A new and larger study is being planned. (In figure 3, I show an artists impression of a cluster of Musgrove windmills.) The results of the first "North Sea Study" look very encouraging; indicating a potentially very large resource (upto 30% of U.K. electricity needs) and costs that could make off-shore wind energy competitive very soon (e.g. an electricity cost of about 4p/kWh.)

You will see then that U.K. plans are starting to progress, which is quite encouraging. Yet it will be some time before we have the first large (megawatt size) U.K. windmill built. Only then will we start to get the <u>hands-on-experience</u> which is so essential!

By contrast the U.S. has already built and had experience with a "first generation" of windmills (Mod. 0 and Mod. 1) and is now completing a triplet of windmills which can be considered as "second generation" in concept and design (Mod. 2), and contracts are now being placed for a "third generation," series of windmills (Mod. 5 and Mod. 6). The specifications of the Mod. 0, 1 and 2 windmills are listed below:

Model	Diameter	Power	Number Built	Builder	Vicinity
Mod. 0	125 ft.	100-200kW	4	Westinghouse	Clayton, Ohio etc.
Mod. 1	200 ft.	2 MW	1	G.E.C.	Boone, N. Carolina.
Mod. 2	300 ft.	2.5 MW	3	Boeing	Seattle, Washington.

More startling than this activity is the very recent bill just passed by Congress to spend approximately $1000M. of government money to subsidise the "commercialisation" of both large and small windmills in the U.S.A. It is against this fast moving U.S. programme that we must examine the relatively slow and ponderous movement of the U.K. programme and wonder whether U.K. industry is likely to be able to compete in future world and even U.K. markets!

So far I have simply given you a review of what is happening (or not happening) on the construction front in the U.K. and elsewhere. I will now attempt to look at some of the technical and engineering questions in wind power. I will also discuss some of the research carried out at Reading University.

1. Fluctuating nature of the wind. (see Refs. 1-7)

The first thing that strikes us about the wind is its fluctuating nature and its relatively poor predictability. This means that the output of a wind turbine may be fluctuating second by second, minute by minute and hour by hour. Each time scale gives us a different set of problems. Examining these problems and establishing whether a strategy can be found for integration of wind energy at national grid level is the subject of a study at Reading which is SRC funded and has been going on for about two years. This work is being carried out by Dr. Peter Musgrove and myself and Dr. Ervin Bossanyi and Mr. Geoffrey Whittle. We have published six papers from this study (see appendix) and I will attempt to review some of the more interesting results.

a) Wind statistics

The power available in the wind at any moment in time increases with the third power of the wind velocity.

$$P \propto V^3$$

This means that if the wind has a turbulence of ± 10% during a given time period then the output power of a windmill will be fluctuating by ± 30%. This unfortunate fact means that the vagaries of the wind are greatly multiplied in wind power systems.

The situation is not quite as simple as this when we start to look at the windmill response characteristics. Let me define a number of parameters which we need to know when defining the performance of a windmill.

Figure 4

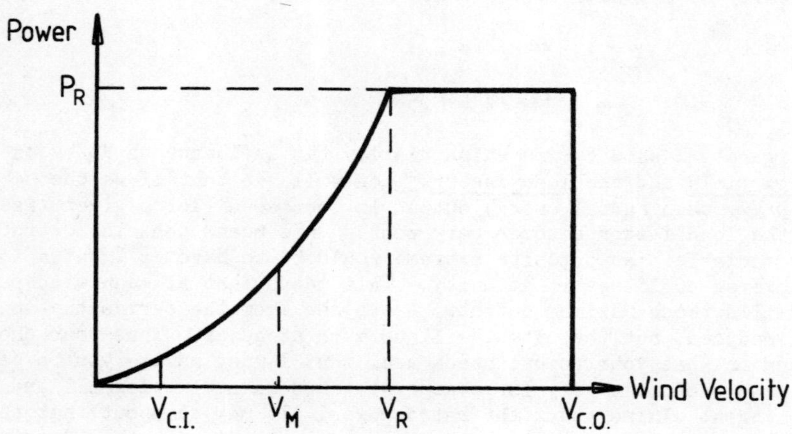

In figure 4, I show a typical windmill response characteristic. This is an idealised curve as I have not taken into account frictional effects.

You will see that the power increases with V^3 upto a ceiling value P_R. P_R is called the rated power (it occurs at wind velocity V_R). The plateauing of the response curve is a deliberate part of our operating strategy, and is obtained by changing the windmill geometry by one means or another. With a horizontal axis machine we may simply change the blade pitch - partially feathering the blades. In the case of the Musgrove vertical axis windmill we tilt the blades away from the vertical. Why do we adopt this strategy?

The point is that one wishes to optimise the engineering costs. By deciding on a maximum output power we then define the strength of the tower blades, drive train and the size of the generator! When the wind exceeds V_R we continue to generate at the rated power P_R. Evidently, we are choosing to throw away some available power on such occasions. But assuming we have rated the machine correctly for the site it would be uneconomic to try and capture the extra energy, as the engineering costs would escalate.

Two further constants are defined namely V_{ci} and V_{co} These being the wind velocities at which we cut the windmill in (start it up) and at which we cut it out (shut it down). The choice of all these parameters is part of the engineering design process for a given windmill at a given site. The choices are very important as they affect many aspects of a windmills operations including the following:

a) Annual energy yield.

b) Effective load factor = annual output/output at rated power.

c) Number of cut-ins per year.

d) Number of cut-outs per year.

e) The general operational flexibility.

Typically we may have $V_R/V_M = 1.4$

$$V_{co}/V_R = 2$$

$$V_{ci}/V_M = 0.5$$

In figure 5 I show curves which display the influence on V_R/V_M on the annual energy yield and the load factor. You will see that if we choose a high value of V_R/V_M, then annual energy output is increased (for a given size windmill) but the load factor becomes very poor. This means that the output would be very variable. An opposite extreme would be to have V_R/V_M at a low value. For example we could set it at unity. This means that at mean windspeed the windmills reach maximum output. As we see from the curves the annual energy is much reduced, but the output will be much steadier. Thus your choice must depend on what your actual needs are. But I must advise you to beware of salesmen who put a very large generator onto a small windmill and then make extravagant claims about the rated power! It may turn out that the wind is seldom strong enough to turn the heavy generator! Ask them to give the ratio V_R/V_M for your site. Sensible values lie in the range 1.2 up to about 2.2.

The next question that we asked in our Reading studies was whether the situation with regard to fluctuating output would look better if one had a cluster of windmills. Dr. Bossanyi showed that for clusters of megawatt size windmills spaced about 500 metres apart there is indeed a considerable improvement as you increase the numbers. This result is shown in figure 6.

Wind power integration into the National Grid (see Refs. 5, 6 & 7)

At Reading we have been developing a quite complex model of the U.K. National Electricity Grid System. Geoff. Whittle has been running this model with a range of different assumptions with regard to both the "conventional plant" and the wind plant. Basically what the model does is to adjust the various types of plant - each within its boundary conditions - to meet the hourly load as efficiently as possible. Into this model Geoff feeds actual hourly wind data and actual load data and he attempts to optimise the system according to chosen control algorithms. This study is only half way through; but already it is yielding some very interesting results! I should mention that similar modelling is being done by Dr. R.A. Taylor and Dr. T. Rockingham of the CEGB Planning Department in London. We keep in very close touch.

In figure 7, a typical day in the life of our model is shown. The inputs into the grid include: wind, fossil fuel steam turbines, gas turbines; and nuclear (which was taken as a steady value and is not plotted to keep the graph less cluttered). We see that on the day in question the wind started at a high value, and then died out rather rapidly during hours 1 to 3. Fossil fuel plant on partial load or in the process of being run-up (total run-up time = 8 hours) was able to make up much of the deficit; but ultimately quite a large amount of gas turbine peaking plant had to be brought on to make up the rest of the deficit.

We have followed load and wind in this manner, hour by hour, for periods of a year. Some of the many statistical results obtained are illustrated in the next two figures.

In figure 8a we see actual annual yield of wind plant and steam plant as a function of installed wind capacity (rated wind plant/maximum system demand). We see that the wind energy curve increases practically linearly. This means that almost all the wind energy is successfully used within our control strategy.

Thus the model is saying that as much as 30% of our electricity needs could be met by wind - within a sensible operating strategy.

Figure 8b tells us that nonetheless, quite a lot of expensive gas turbine fuel would have to be burnt, within a wind strategy, in order to make good wind power defecits (of the sort we discussed on a one day basis). The gas turbine annual energy yield required increases a little faster than linearly with the installed wind capacity.

The next question we asked is how serious will be the cost of operating back-up plant in order to make up short-falls of wind-power. Fuel costs for gas turbines are two to three times as high as coal-steam plant fuel costs. So we asked whether it might be cheaper to adopt a different back-up strategy in which we would keep a lot of steam plant on partial load in order to take-up fluctuations in wind power! Such a plan also sounds expensive and inefficient. In practice we examined a range of strategems ranging from back-up relying on partially loaded steam plant to back-up relying entirely on peaking plant. Our findings are shown in figure 9b. Relative fuel cost saving (for windpower) is plotted against the amount of spinning reserve (in a given operating strategy). A relative fuel cost saving of 1.0 would mean that the back-up strategy was causing no waste of fuel. In practice we see an optimum at 60% reliance on spinning reserve. At optimum the fuel cost wasted on back-up requirements (to wind) is about 12%.

The conclusion is that about 12% of the energy yielded by the wind generation plant is in fact lost because of the necessary back-up strategy. I consider this to be a very important result: - the penalty does not look to us to be too damaging for wind energy. The top two graphs 9a and 9b simply show the steam plant effective operating efficiency falling off (for the reasons just given), as, we introduce more wind plant, or demand more hot spinning reserve.

Similar studies might be made with regard to the introduction of more nuclear generating plant into the U.K. Nuclear plant is not normally considered for rapid load following. Hence greater nuclear capacity on the grid would imply more difficult operating conditions for the steam and peaking plant, with a consequent worsening of operating efficiency - quite similar to the wind situation I have just described. I am not sure the pro-nuclear lobby have properly assessed the penalties - as we have attempted to do for wind.

Small systems study

About one year ago we decided to extend our systems studies to look at the integration problems for wind generators into small power systems. We had in mind communities who were isolated from large electricity grid systems. We had in mind Scottish Island communities and also Third World situations. Barbara Sexon and Graham Slack have worked on this project for under a year. And yet already they have produced a number of very useful insights into the problem. Our starting point was to define a system which included wind power plus diesel plus a possible storage medium. The questions we asked is can we define a system and an operating strategy which does not force the diesel plant into inefficient or inappropriate operating conditions. But let me say no more! I do not intend to steal Ms.Sexon's thunder. Let me only add that we hope quite soon to translate our work from modelling studies to an actual practical exercise with hardware.

Wind energy hardware activities of the Reading Group. (Refs. 9, 10)

Upto this point I have emphasised the paper studies going on at Reading University. We also have a number of interesting hardware projects and activities at Reading, some of which I will enumerate.

a) Field measurements collaboration (see Ref. 13)

We are members of this collaboration which is co-ordinated by Dr. Don Swift-Hook of the CERL's Leatherhead Laboratory. Some six universities (Exeter, Cambridge, Swansea, Reading, Cranfield, Newcastle), two industrial firms (P.I. Engineering and Sir Henry Lawson-Tancred) and one Government Laboratory (Appleton-Rutherford) are involved in the project. The purpose is to make assessments of seven small U.K. windmills and to establish uniform standards of measurement. The list of windmills is as follows:

Site	Windmill type	Diameter
Exeter	Vertical axis (Cyclic pitch)	5m
Swansea	Horizontal axis(3 Bladed)	5m and later 15m
Newcastle	Vertical axis (Darrieus)	5m
Cambridge	Horizontal axis	3m
Appleton-Rutherford	Vertical axis (Musgrove)	6m
P.I.	Vertical axis (Musgrove)	6m
Lawson-Tancred	Horizontal axis (3 Bladed)	17m

The Reading group and CERL, Leatherhead have jointly set up a mobile laboratory, with data logging facilities, meteorological towers and measuring equipment. We propose to measure the following properties:-

(i) Windmill performance under various conditions of turbulence. In this programme windmill yaw response will be one special feature studied.

(ii) Windmill wake in the near field and far field. We need to know how much wind energy is removed from a "wind front" by a row of windmills and hence how much energy will be received by later rows. This study is of great significance for the planning of wind clusters on land or in the sea.

(iii) Studies of materials and vibrations. A new section to our enterprise has recently been set up by Dr. Tony Pretlove of Reading and Dr. Peter Worthington of CERL. They intend to make experimental studies of structural and vibrational problems in the windmills within our programme.

Let me note that the scheme for setting up a mobile laboratory which will help to bring together the researches from seven small windmill sites is an ambitious and visionary idea. But it is also full of headaches; as the co-ordinating problems are pretty formidable. Yet I believe that we can make it work! But in the long term I believe that U.K. wind energy would greatly benefit from the setting up of a central test site to which commercial windmills would be brought for testing and <u>validation</u>. Such an idea is now under discussion. You might like to let me know if you see it as helpful to you. The site would also act as a show place where potential customers would come to look at and discuss the machines with independent experts.

b) <u>Commercialisation of the Musgrove vertical axis windmill</u>

Dr. Musgrove has given help and advice to P.I. Engineering who are now marketing a self-starting version of his vertical axis-variable geometry windmill. He is now advising a group lead by McAlpine who have recently received D.En. funding to design a 25 metre version of his machine. A six metre version of the Musgrove windmill has been built and erected by Mr. A.R. Mortimer of SRC on the Appleton-Rutherford Laboratory site. This machine is being used by Peter Musgrove and Mr. Giles Stacey of Reading in studies of windmill performance (Ref. 8) and (Ref.12) operational characteristics. Drs. Pretlove and Worthington are also carrying out strain and vibration measurements on this machine. It may also be used by McAlpine in the D.En. programme.

c) Third World Applications of Windmills

Professor Dunn is responsible for this aspect of the work. He is energy consultant to ITDG which has been developing a windmill for water pumping applications, on the Reading site. This machine is now in the process of being taken up by local entrepeneurs in a number of Third World countries. A particular problem is to develop a water pump well matched to a windmill drive. Prof. Dunn and colleagues have been working on a number of pump types.

Of particular interest is a "kinetic pump" invented by Dr. John Burton, a visiting Fellow to the Reading group. His pump attempts to match the V^3 response characteristic of windmills (V = wind velocity). If successfully applied it will typically double the annual water through-put for any given size of windmill, with respect to conventional piston pumps. This is an exciting breakthrough if this pump can be taken through to successful commercialisation!

Mr. Tayib Eisa, a Sudanese Graduate Student has recently carried out performance measurements on the ITDG windmill. He hopes to return shortly to the Sudan to become their expert in wind energy.

I have presented a rather wide survey of the U.K. wind energy scene. I have tried to give you some feeling for what is going on commercially and politically as well as discussing briefly some technical points. For further details of the Reading research programme you should consult the list of papers. The best plan is to contact the first named authors in most cases he/she has worked most closely with the points of detail. To conclude with, how I view future prospects for wind energy in the U.K. I believe that wind is now looking very promising and I believe that we are now ready to start commercial exploitation. But I consider it essential that the U.K. government makes funds available (just as the U.S. government is doing) to help U.K. industry and customers alike over the initial commercialisation hump.

I would like to thank Miss Joy Clancy for her help in the preparation of this paper. Miss Clancy is the Industrial Liason Officer of the Reading Energy Group. She is the person whom you can contact to discuss aspects of the Reading programme.

Fig.1. Members of the Reading Energy Group Involved in Wind Energy Research

Department of Engineering

Prof. P D Dunn
Prof N H Lipman
Dr P Musgrove
Dr E Bossanyi
Dr D Svendsen
Mr G Whittle
Mr G Stacey
Mr E Fordham
Mr J Turner
Ms B Sexon
Mr G Slack
Mr E Preston
Mr T Eisa
Dr A J Pretlove

Department of Meteorology

Dr J R Milford
Dr A Ibbetson

I.T.D.G.

Mr P Fraenkel
Mr M Ewens

Appleton and Rutherford Laboratory

Mr J Halliday

Fig. 2. The MOD-1 (2 MW) wind turbine installed in Boone, North Carolina.

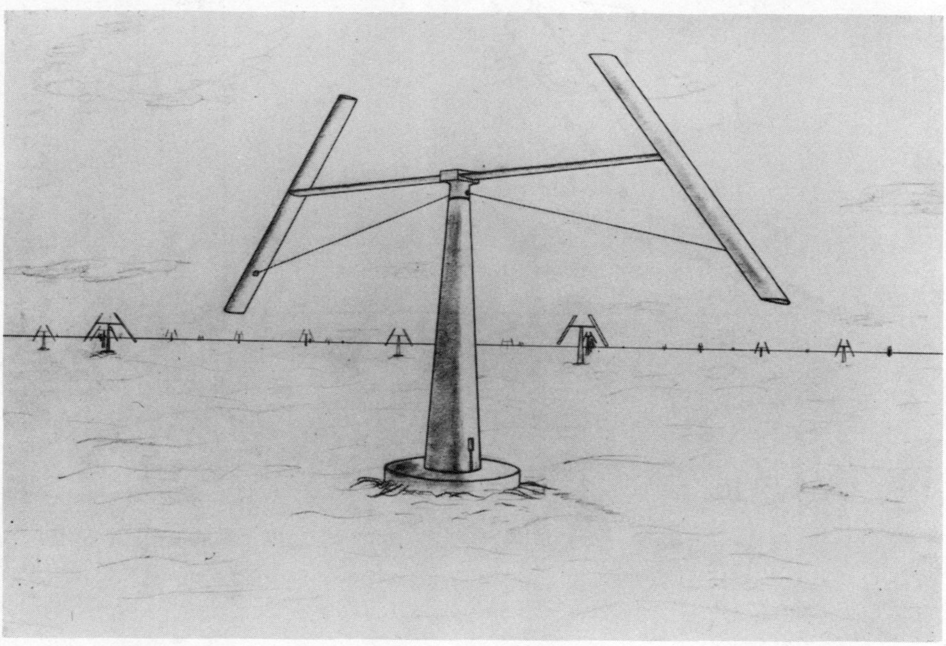

Fig. 3. Artist's impression of cluster of Dr. Peter Musgrove's windmills.

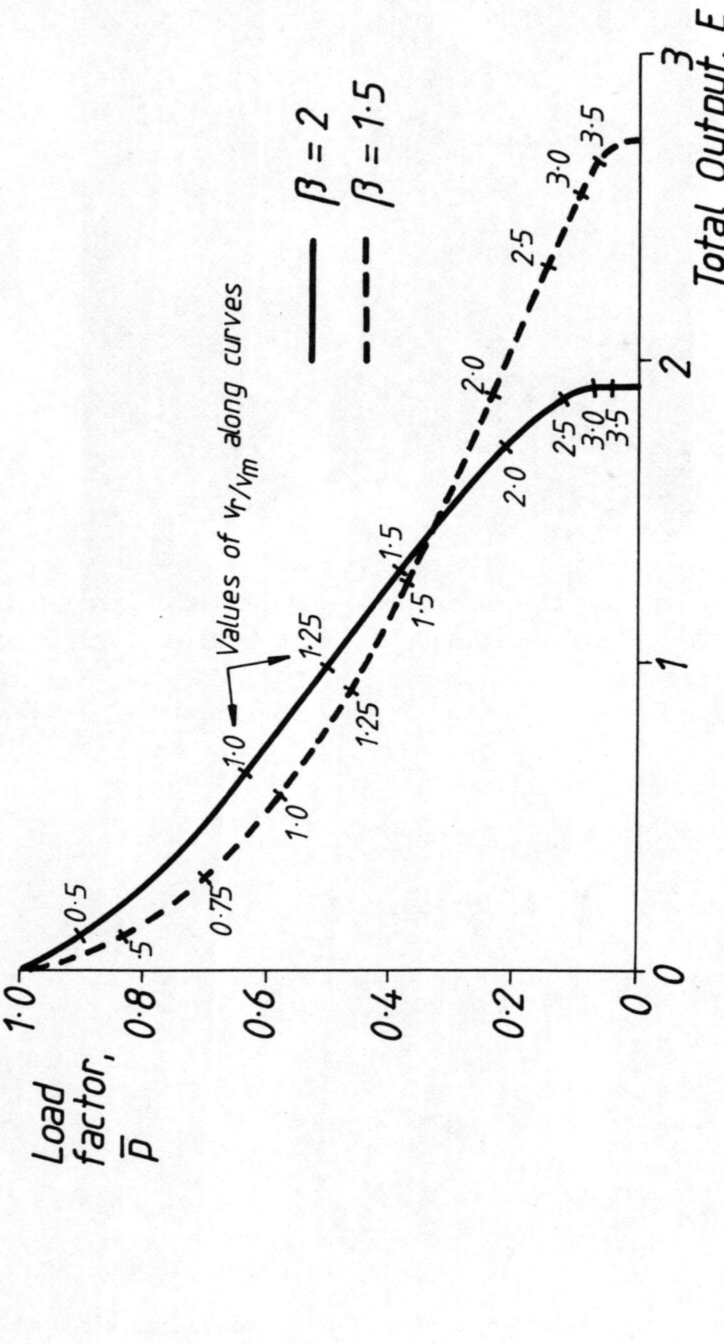

Figure 5. Load factor vs. Total Output

Figure 6: Probability Distribution of Power Output

when $\overline{V} = 0{\cdot}75.V_{rated}$, with 20 % turbulence

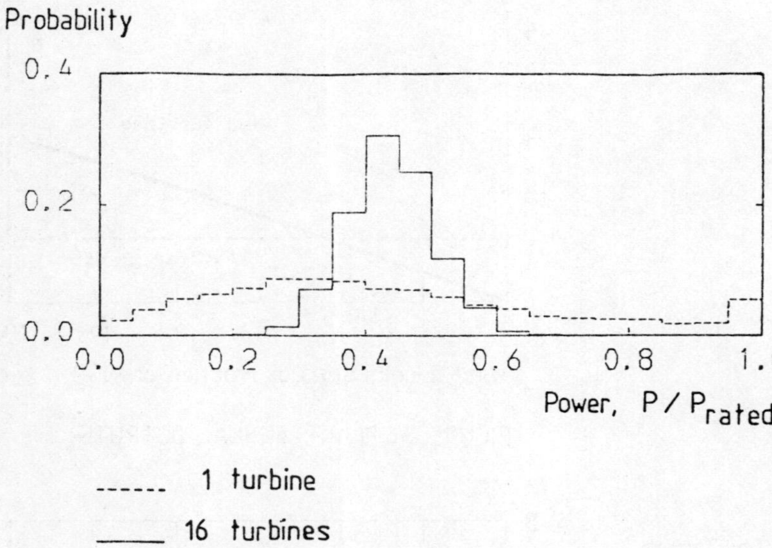

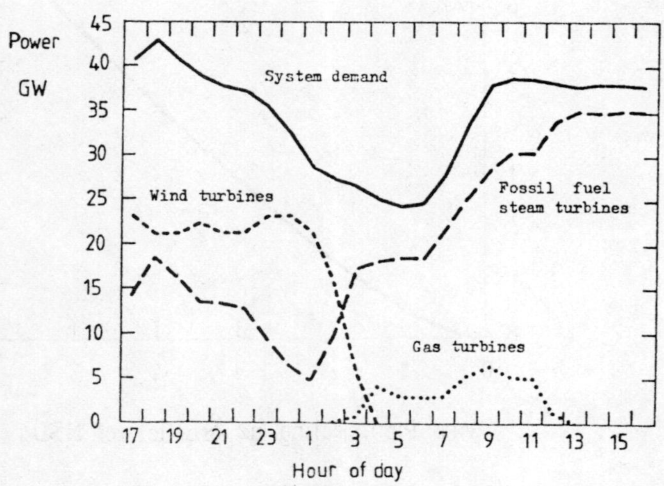

FIGURE 7 PLANT HOURLY OUTPUTS (24 Hour Period)

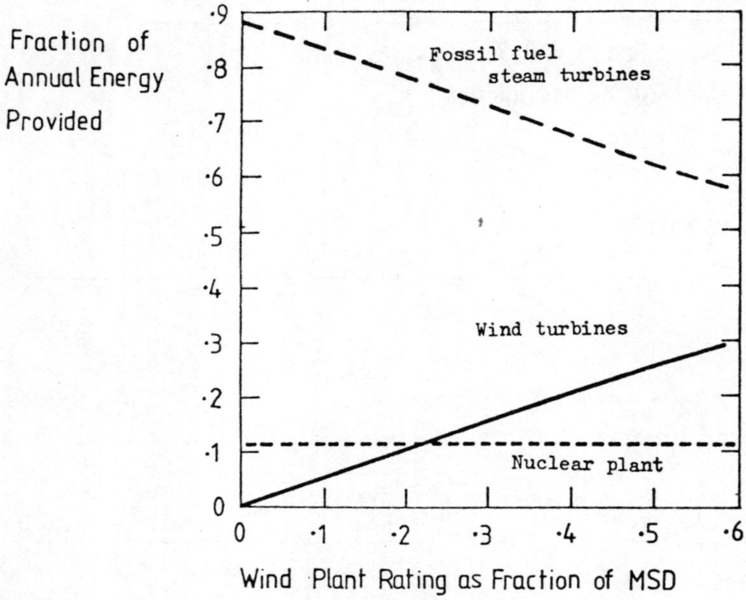

FIGURE 8a PLANT ANNUAL OUTPUTS

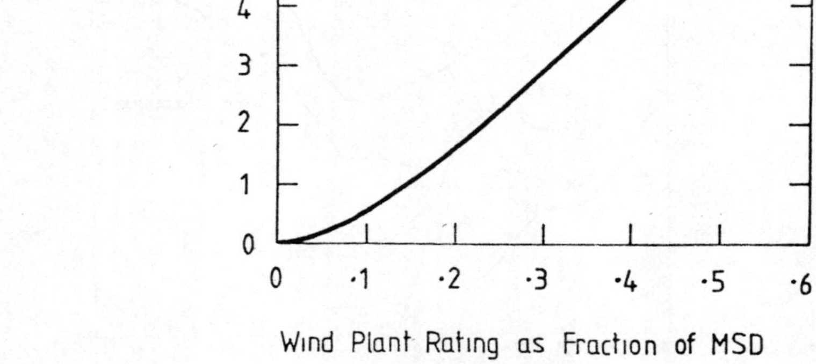

FIGURE 8b GAS TURBINE PLANT ANNUAL OUTPUT

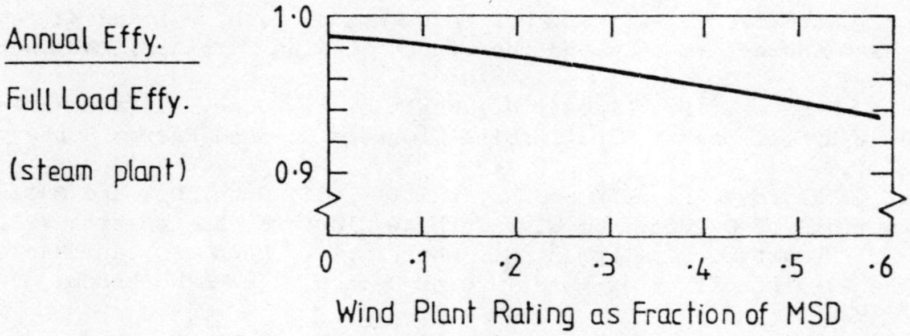

FIGURE 9a EFFECT OF WIND PLANT RATING ON. EFFICIENCY

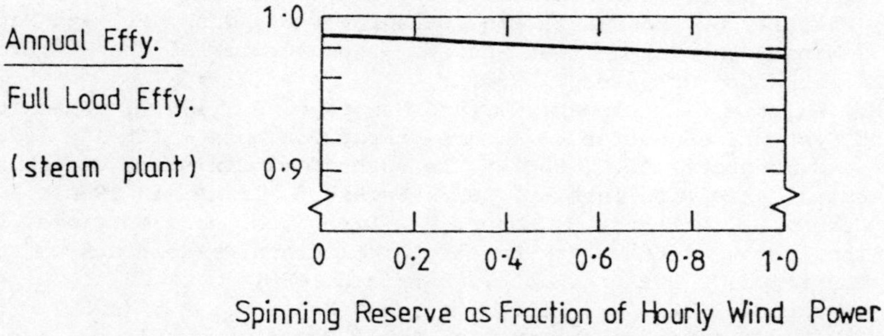

FIGURE 9b EFFECT OF SPINNING RESERVE ON EFFICIENCY

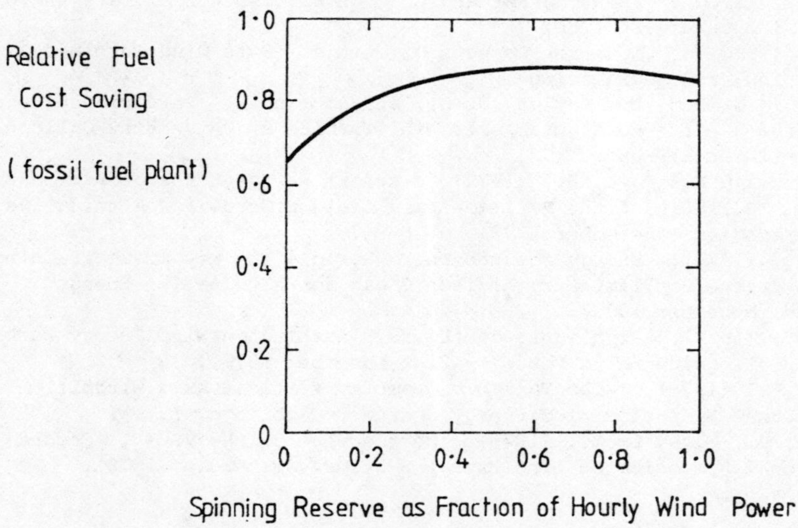

FIGURE 9c RELATIVE FUEL COST SAVING

REFERENCES

1. Bossanyi, E., Musgrove P., Lipman, N., Whittle, G., Dunn, P., and Maclean, C., (1979). Wind Characteristics and the Output of Wind Turbines, BWEA Workshop, Cranfield, 1979.
2. Bossanyi, E., Musgrove, P., Lipman, N., Whittle, G., Dunn, P., and Maclean, C., (1980). The Efficiency of Wind Turbine Clusters. Wind Energy Conference, Copenhagen, 1980.
3. Bossanyi, E., Musgrove, P., Lipman, N., Whittle, G., Dunn, P., and Maclean, C. Fluctuations of Output from Wind Turbine Clusters. Engineering Vol. 4.1.
4. Bossanyi, E., Musgrove, P., Lipman, N., Whittle, G., Dunn, P., and Maclean, C. (1980). The Predictability of Wind Turbine Output. (BWEA Workshop, Cranfield, 1980).
5. Whittle, G., Bossanyi, E., Musgrove, P., Lipman, N., Dunn, P., and Maclean, C., (1980). Simulation Model of an Electricity Generating System Incorporating Wind Turbine Plant. BWEA Workshop, Copenhagen (1980).
6. Lipman, N., Whittle, G., Bossanyi, Esq., Musgrove, P., Dunn, P., and Maclean, C.,(L980). Wind Power Integration Studies - An overview of the R.U.Programme. BWEA Workshop, Cranfield, 1980.
7. Bossanyi, E., Whittle, G., Lipman, N., and Musgrove, P. (1981). Wind Turbine Response and System Integration. (Future Energy Concepts - IEE Jan 1981).
8. Stacey, G., and Musgrove, P., (1980). The Rutherford Six Metre Variable Geometry Vertical Axis Wind Turbine. (BWEA Workshop, Cranfield 1980).
9. Slack, G., Sexon, B., Collins, R., Dunn, P., Lipman, N., and Musgrove, P., (1980). Wind Energy Systems with Battery Storage and Diesel Back-up for isolated Communities. (BWEA Workshop, Cranfield 1980).
10. Sexon, B., Slack, G., Dunn, P., Lipman, N., and Musgrove, P., (1980). Wind Turbine/Diesel Electricity Generation for Small Remote Communities. (Report for U.N. - Private).
11. Lipman, N., (1980). Blowing Hot and Cold on British Windmills.(In The Guardian of 28th August 1980).
12. Stacey, G (1980). Vibration Aspects of Vertical Exis Wind Turbines. (To be published in "Wind Engineering").
13. Turber, J., Fordham, E., Taylor, Dr R., and Lipman, N. Performance Measurements on the 5 Metre Horizontal Axis Wind Turbine at University College, Swansea. (Internal Report).
14. Lipman, N., and Musgrove, P., (1979). Report on the Washington Wind Energy Workshop. (Bulletin of the Parliamentary Liaison Group for Alternative Energy Strategies - December 1979).
15. Lipman, N., (1979). Energy for the Third World - The Key Issue Training. (Bulletin of the Parliamentary Liaison Group for Alternative Energy Strategies, December 1979).
16. Lipman, N., (1979). Highlights of the 1979 Washington Wind Energy Workshop. (BWEA/IEE Colloquium "Wind Power" - 21st November 1979).
17. Musgrove, P.? (1976). The Variable Geometry Vertical Axis Windmill. (International Symposium Wind Energy Systems, Cambridge, 1976).
18. Musgrove, P., Mays, Dr.I., (1978). Development of the VGVAW. Second International Symposium on Wind Energy Systems, Amsterdam, 1978).

Windpower for Agriculture

Andrew Stobart

Director, Trimble Windmills, 55 East Parade, Harrogate,
North Yorkshire HG1 5LQ, UK

ABSTRACT

Windpower was much used in agriculture in the past, and could be again to provide an independant, inflation proof energy source, allowing conserving of fossil sources. Its cost effectiveness depends on the close matching of the using equipment with the windpower source. Larger energy supplies might be better met with a number of small units, rather than one large one. The most cost effective uses for windpower on a farm are heating, and water pumping. Wind is a "crop".

KEYWORDS

WINDPOWER, HEATING, PIG REARING, AGRICULTURE, FUEL CONSERVATION COUNTERING INFLATION, WATER SUPPLIES, DISTRICT HEATING, FARM INCOME.

The earliest use of windpower in England is associated with agriculture with one Nicholas renting a mill from the Knights Templars in the village of Weedly, near Hull in Yorkshire, c.1182. He almost certainly ground the corn for the village; his father, one Gamel, renting a conventional water mill nearby, also providing the only mechanical power source for that community. In the 1300's the rent roll of Ford Castle in Northumberland put the field rents at some £5 per annum, and the mill rent at £15, showing the value then of running the only power source, other than man and beast.

The concern with a power source continued through the centuries, a manual on Scottish agricultural engineering published in 1814 showed a thrashing machine with a drive from a windmill, a water mill, or, if wind and water failed, six horses. This brings into sharp focus the two problems of using windpower, its variability of supply, and the need for a back up, or storage facility. In the traditional milling role, the storage was that of the grain to be milled, and the flour after milling. One had both a ready supply of work, and a storage of product. The same principles applied to the water pumping or "tin" windmill, once common throughout the UK, and still used in some areas, there are several for instance on the coast North of Aberdeen. The water to be pumped was stored in the ground, and a storage tank on the farm held a sufficent supply to cover windless periods. The windmill worked as it could, transferring the water from one "storage" to another. It is likely that in these days of high mains water charges the re-introduction of windpumps would be economic in many places.

But a modern farm has many more requirements. Tractor fuel apart, the farm has an energy input in electricity, coal, gas, and diesel, as well as fertilisers. Where the fuel is at present used for static applications, including running electric motors one can consider wind power. But the simplest and cheapest application of windpower is for heating, particularly where the heating is at present done with electricity. Pig house heating, potatoe storage, mushroom growing, greenhouses, dairy hot water, and bottlewashing hot water, to name but a few. In general the heating can be done without the use of expensive batteries, and as the heat requirement of a building increases in line with increasing wind speeds, and so does the heat produced from a windmill there is a convenient "balancing" of supply and demand. Fig. I shows the output from a Trimblemill in relation to wind speed, Fig. II a typical layout of a water heating unit attached to a circulating hot water system, and Fig. III a diagram of the electrical control.

For use with electric motors, some form of battery storage is required. Preferably the equipment run from the batteries should take current direct, in the form of d.c. power, this removes the need for inverters or converters to turn d.c. power into a.c. This is particularly important for equipment with a high starting current such as a deep freeze. The inverter needed to start it is some four times as large as that required to run it once started. Direct connection to batteries however can produce the required starting current with no trouble. Also the more equipment that is inserted between the Trimblemill and the final using equipment the greater is the quantity of energy which is dissipated as heat to little or no useful purpose. Thus reducing the eventual "income" from the unit, and adding to the overall capital cost of the whole installation.

In certain instances it is technically possible to "float" on the mains supply, and this practise is not illegal in Scotland, and is in operation with a number of Hydro Electric installations in private ownership. This means that all the current produced by the Trimblemill would either be used on the farm, or exported to the grid, and credited against future supplies of grid power. However this practise has not yet been agreed with the Hydro Board. They are currently conducting Windmill trials with a view to assessing the problems involved, and finding ways of dealing with them. The ESB in Ireland are conducting similar trials with a "simulated" engine driven alternator.

For larger power requirements, more than one alternator can be mounted on one tower, or several machines erected at suitable sites around the farm and the power combined. This could be preferable to installing one large machine, from the viewpoint of protection from total equipment failure. Also smaller machines, in that they have no gearbox, are in general more efficient collectors of energy than the larger units on an income to cost basis. Desk studies indicate the probable maximum size of unit, using existing Trimblemill technology, as being about 120 KVA, three phase, 440 volts, at a wind speed of 22 mph, (10 m/sec), and with a maximum output of around 300 KVA. This size of unit could be applied to district heating schemes for private housing, heating factories etc., as well as possibly providing power to large agricultural energy users.

Why use wind power? Like the man said of the mountain, "its there". And as a form of energy that is roughly constant in yearly income in electrical terms, it is an inflation proof (and tax free) source of energy, given a suitable site, and method of use. It is also worth noting that on the "rated" output of a Trimblemill, 5kW, its cost is not much more per kW than that of a nuclear power station. And the latter requires fuel, operating labour, waste disposal. For the larger "arrays" mentioned above, the installed cost of windpower units could well be significantly cheaper than nuclear power. Equipment life could also be longer than the twenty or so years predicted for the nuclear units. Figs. IV and V show a Trimblemill.

Another reason for using windpower is to enable the farmer, and other countryside users to once again contribute to their own, and others energy requirements. And to provide a dispersed, and quickly expandable source of energy for agriculture as a protection against interruption of the more conventional sources. Most farming in the British Isles is a net importer of energy. The output in food being less, in energy terms, than the input of fuel. Effectively we are at present "eating oil".

A sudden interruption of oil supplies could have grave consequences for agriculture in general, so if its demand had been lessened by the installation of alternative supplies it would be more able to "weather the storm". Even a return to horsepower would take quite a few years, wind powered equipment can be installed in a matter of weeks. It also produces a "cash flow" that much faster.

A third reason is that wind could provide a farmer with an additional income. If the national wind energy policy favoured the installation of small to medium sized machines as being considered in Orkney and Ireland, then the siting of equipment might prove less difficult than recent press articles suggest. Wind energy could be considered as a "crop" partly for on farm consumption, with an exportable surplus.

It is likely that other uses of windpower than those discussed above could be made available in the long term. This will depend on the relative cost of the windpower, compared to the cost of alternatives. But the user who installs windpower, or any other form of non fuel power now, is that much better placed to make increasing use of it as the years go on. It is surely no coincidence that the country with the strongest financial system in Europe, and possibly in the world, Switzerland, relies very largely on power produced from non fossil sources. The Hydro Board installations here in Scotland, once only marginally cost effective, now produce some of the cheapest power available. The farmer with the windpump does not face increasing charges each year.

The typical view of the farmer is of a man of sturdy independance. This position has been sadly eroded recently, but as the field, if farmed correctly, produces an income and thus independance, so can the wind through energy farming.

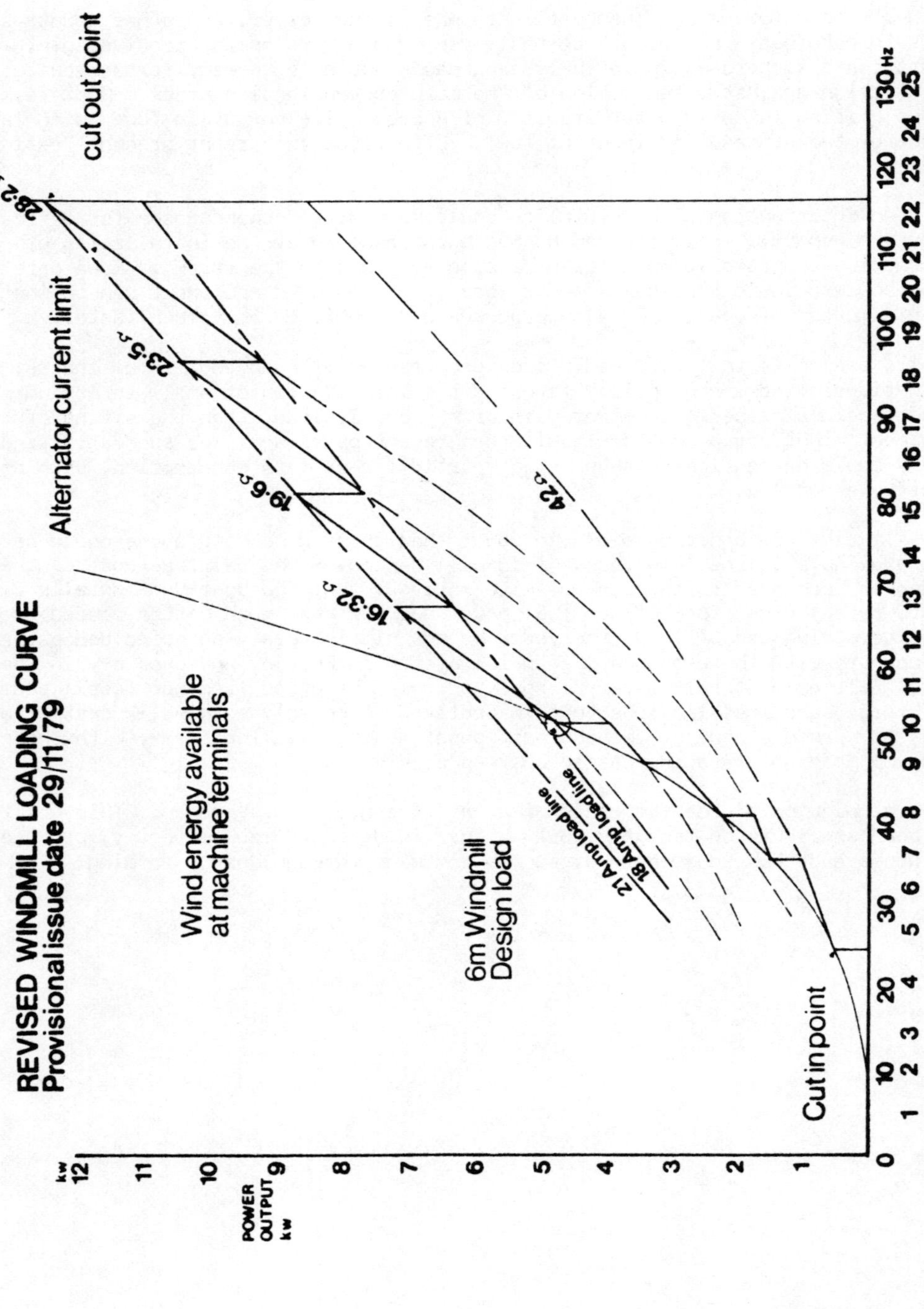
REVISED WINDMILL LOADING CURVE
Provisional issue date 29/11/79
Wind energy available at machine terminals
Alternator current limit
cut out point
6m Windmill Design load
Cut in point
18 Amp load line
21 Amp load line
28·2 Ω
23·5 Ω
19·6 Ω
16·32 Ω
42 Ω
POWER OUTPUT kw
kw
Wind Speed m/s
Hz

Fig. I

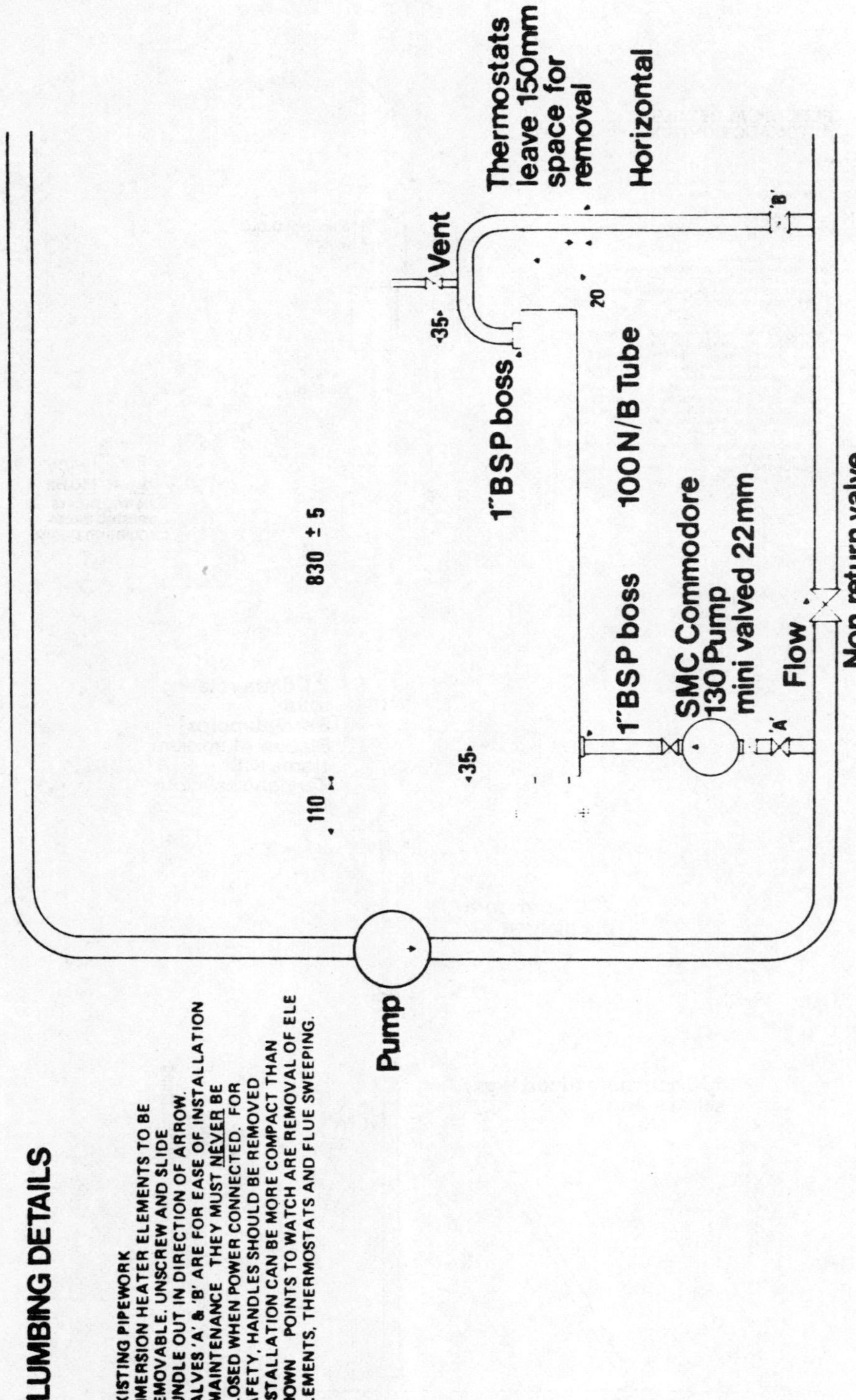
PLUMBING DETAILS
1. EXISTING PIPEWORK
2. IMMERSION HEATER ELEMENTS TO BE REMOVABLE. UNSCREW AND SLIDE BUNDLE OUT IN DIRECTION OF ARROW.
3. VALVES 'A' & 'B' ARE FOR EASE OF INSTALLATION + MAINTENANCE. THEY MUST NEVER BE CLOSED WHEN POWER CONNECTED. FOR SAFETY, HANDLES SHOULD BE REMOVED
4. INSTALLATION CAN BE MORE COMPACT THAN SHOWN. POINTS TO WATCH ARE REMOVAL OF ELE ELEMENTS, THERMOSTATS AND FLUE SWEEPING.
Pump
110
35
830 ± 5
35
Vent
20
1"BSP boss
1"BSP boss
100 N/B Tube
Thermostats leave 150mm space for removal
Horizontal
SMC Commodore 130 Pump mini valved 22mm
Flow
'A'
'B'
Non return valve

Fig. II

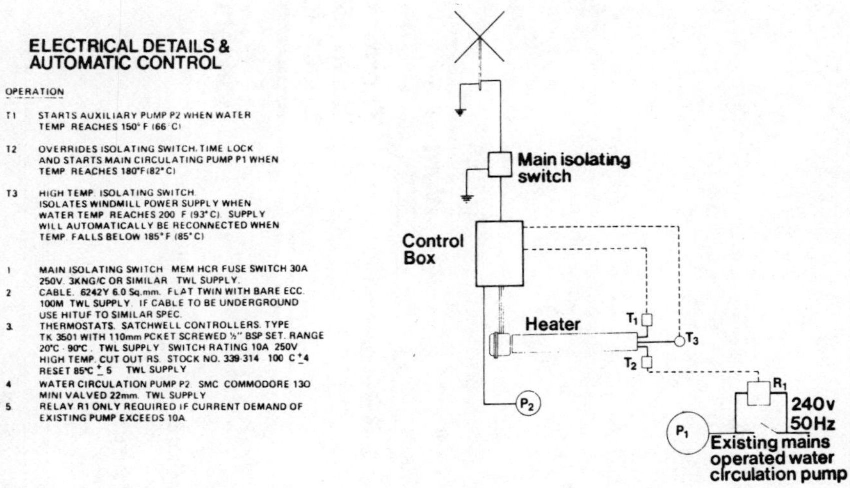

Fig. III

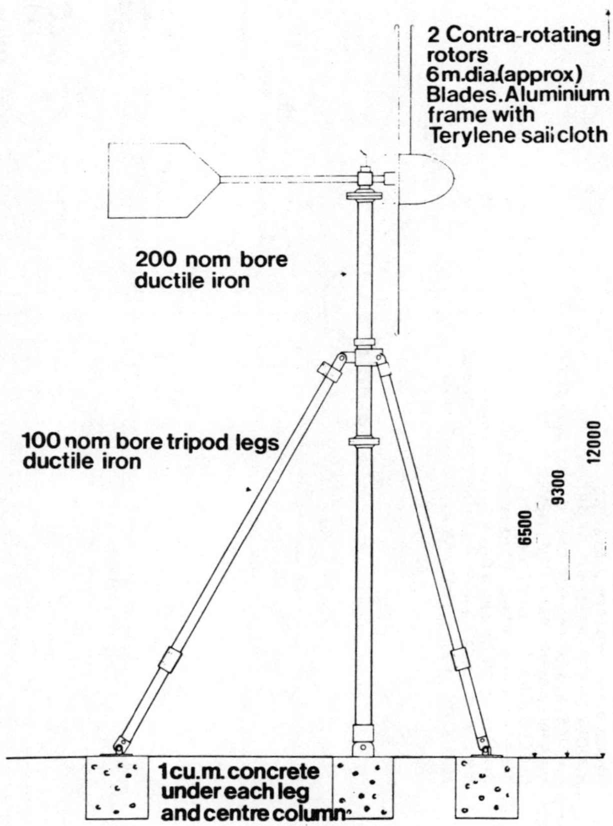

Fig. IV

Fig. V

The Development of a Straight-bladed Vertical-axis Wind Turbine

W. S. Bannister* and S. Gair**

*Department of Mechanical Engineering,
Napier College of Commerce and Technology, Edinburgh
**Department of Electrical and Electronic Engineering,
Napier College of Commerce and Technology, Edinburgh

ABSTRACT

A straight-bladed vertical-axis wind turbine has been constructed at Napier College. The installation has been designed as a test bed in order to test various blade configurations and electrical generator load systems. The turbine has been placed on a test site and will be tested over the next two years. Available evidence has shown that the line of investigation should include a low aspect ratio, high solidity configuration and electrical power take-off to naturally match the turbine and supply a variety of load requirements.

KEYWORDS

Vertical-axis; straight-bladed; high solidity; low aspect ratio; tip-speed ratio; low Reynolds number; induction generator; turbogenerators; wind-power plants.

NOMENCLATURE

a	the slope of the lift curve, i.e. $\frac{dC_L}{d\alpha}$
A	the aspect ratio, $\frac{b}{c}$,
b	the blade span,
c	chord length of blade section,
C_D	blade element drag coefficient,
C_{DO}	blade element drag coefficient at zero incidence,
C_L	blade element lift coefficient
C_{Lmax}	maximum blade element lift coefficient
C_p	wind turbine power coefficient (i.e. $\frac{P}{\frac{1}{2}\rho bc V^3}$)
n	number of blades
N	rotational speed in rev/min
P	wind turbine power output
r_o	radius of turbine, i.e. distance from axis of rotation to blade
R_e	Reynolds Number
T	torque about the wind turbine axis
V	wind velocity well upstream of the wind turbine
β	tip speed ratio $\frac{r_o\omega}{V}$
ρ	air density
σ	wind turbine solidity, ($\frac{nc}{2r_o}$)
ω	rotational speed of the turbine

INTRODUCTION

Vertical axis wind turbines have been developed over recent years with a number of large Darrieus type turbines being built in America and with the Musgrove turbine being tested in England. Both types of turbine are now well known in the field of wind energy research.

At Napier College, work on wind energy has in the past, centred around student projects. The work on wind energy is complementary to the research work which is being carried out on solar energy. In 1978 work was started on a vertical-axis turbine. It was felt that there were possible alternatives to the basic Musgrove design. The alternative ideas were presented at the 1st Wind Energy Workshop of the British Wind Energy Association, in 1979. Similar ideas, for example, of increasing the solidity to improve performance, were arrived at quite independently by Musgrove and Mays (1978) and were subsequently developed to include the effects of low aspect ratio (1980). Small vertical axis wind turbines are not yet commercially available in any great number. Further work is required to improve the performance of the wind turbine, the electrical power conversion, reliability and overall economic viability. To this end, staff at Napier College are embarking on a research programme, funded by the Science Research Council, which will examine various possibilities which have been highlighted in recent investigations.

TEST FACILITY

An installation which can be used as a test bed for carrying out experiments on various configurations of vertical axis wind turbines has been built. The turbine is now in position on a test site. The basic turbine is shown in Fig. 1.

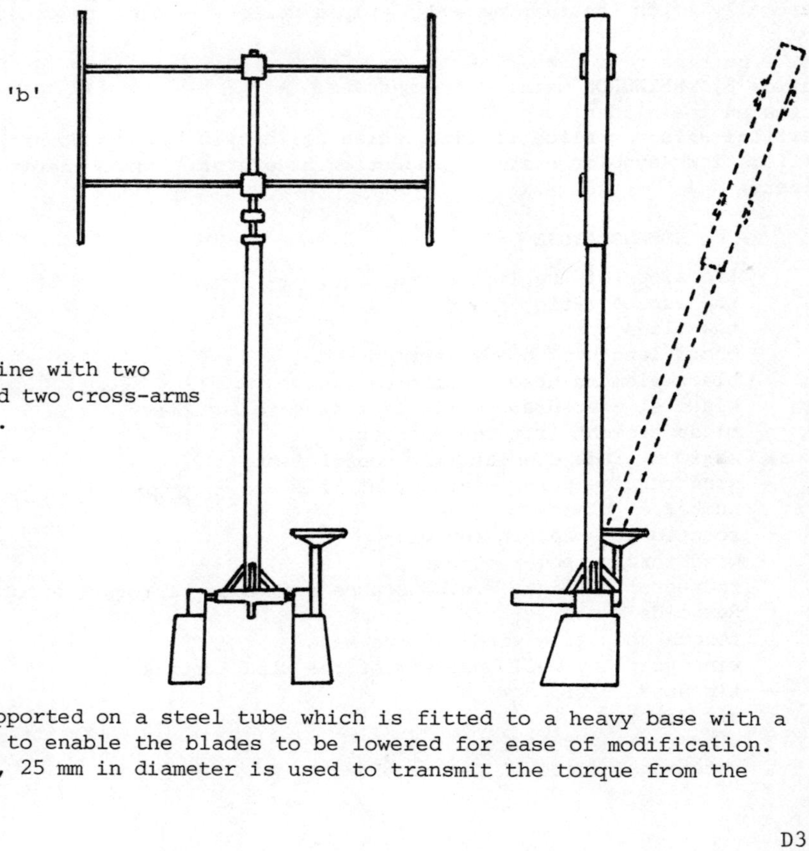

Fig. 1. Wind Turbine with two blades and two cross-arms per blade.

The turbine is supported on a steel tube which is fitted to a heavy base with a tilting mechanism to enable the blades to be lowered for ease of modification. A central spindle, 25 mm in diameter is used to transmit the torque from the turbine to the base of the vertical tube. The turbine us shown with two vertical blades, 2 m long with a constant chord of 150 mm and at a radius of 1.5 m. Electrical power take-off is facilitated at ground level.

The installation is purely a test facility to be used to determine the characteristics of various wind turbine configurations and electrical generator types. Instrumentation is being fitted which will enable measurement of angular velocity, angular acceleration and wind velocity to be made. Strain gauges will also be fitted to the blades to check the blade loading.

DESIGN CONSIDERATIONS

The performance of a wind turbine is normally shown as a plot of power coefficient Cp, against tip speed ratio β as shown in Fig. 2.

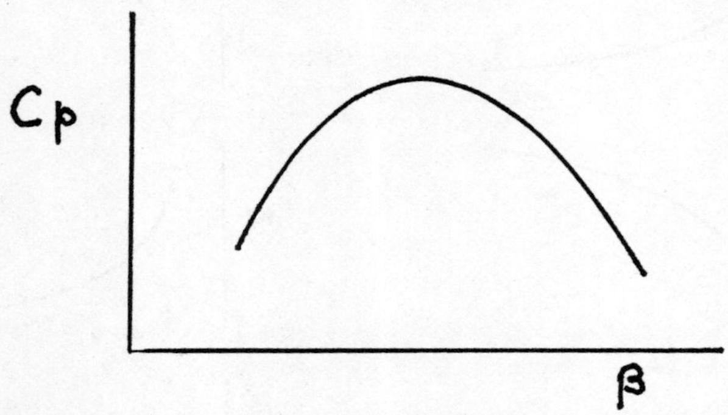

Fig. 2. Typical Power Coefficient Versus Speed Ratio.

For a turbine to run at maximum efficiency, that is maximum Cp, the tip speed ratio, β, should be maintained constant. At very high wind speed, the centrifugal force on the blades become excessive hence there is a need for a rotational speed limiting device. Also, if the turbine has an inherent high cut-in speed, there will be a need for a starting device. The connected load torque on the wind turbine should be arranged to vary as the square of the shaft rotational speed. The object then of wind turbine research is to maintain a high efficiency throughout the wind velocity range, prevent blade failure and at the same time, drive a suitable device with matching characteristics.

AVAILABLE EVIDENCE

1. Blade Section. Originally, the NACA0012 was felt to be the best section, but more recent evidence has tended to suggest that the NACA0015 is more suitable. There is also evidence to suggest that even thicker sections will be better at low Reynolds numbers. (Healy, 1978).

2. Solidity and Aspect Ratio. For a given aerofoil section and with known characteristics, (i.e. C_{Lmax}, a, C_{DO}), is it possible to predict the optimum solidity σ and corresponding β which will give the maximum possible Cp ? Using information from Musgrove and Mays (1978, 1980), Bannister (1979) and Miller (1980), the variations of σ, β and C_{pmax} have been plotted against aspect ratio A in Fig. 3. The scales have been deliberately omitted from these graphs as the calculations assume constant values of a, C_{Lmax} and C_{DO} and no variation with Reynolds number. It is therefore felt that the trends will be correct but the absolute values may be unreliable. Certainly the values of solidity (from about 0.2 to 0.7) are much higher than were initially thought to be the optimum (about 0.1).

2. (contd)
At the higher solidities, the cross-arm drag, which is approximately proportional to β^3, has decreased and as β has decreased as well, the rotational speed and hence centrifugal forces have been reduced to the point where it may be possible to use only one cross-arm per blade. At any fixed wind velocity, the centrifugal force on a blade is proportional to $\frac{\beta^2}{2r_o}$. Also remember that the solidity $\sigma = \frac{nc}{2r_o}$ then for a given chord c and at a fixed wind velocity V it is possible to plot the variations of r_o, N and $\frac{\beta^2}{2r_o}$ against aspect ratio A as shown in Fig. 4. Again this is for maximum Cp.

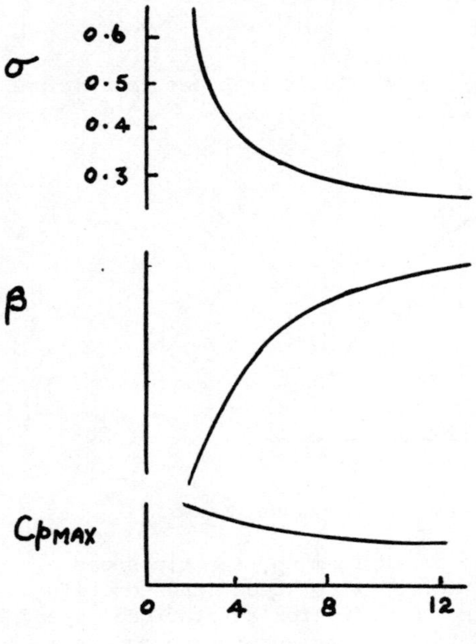

Fig. 3. Predicted variation of σ, β and C_{pmax} with aspect ratio for maximum C_p

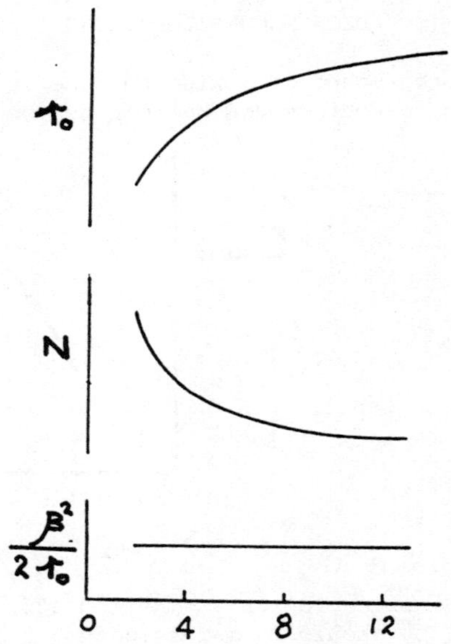

Fig. 4. Predicted variation of r_o, N, $\frac{\beta^2}{2r_o}$ against aspect ratio for a fixed chord.

For a fixed chord c, the centrifugal force ($\alpha \frac{\beta^2}{2r_o}$) may be reduced simply by increasing the number of blades.

Alternatively, the centrifugal force may be reduced by using a blade of larger chord size. Blade cost must be borne in mind and it would appear at present that either 3 or 4 blades must be used. As there is a cyclic variation of torque with one blade, there is the obvious smoothing of the torque output with 3 or 4 blades as well as the other advantages.

3. Blade Material. A large number of wind turbines have wooden blades. Various materials are being used on American turbines with quite a range of methods of construction. It is essential, for small turbines, that the blades be simple to make, easily reproducable in quantity, and cheap. These criteria are not easy to satisfy. At present, we are working on glass reinforced epoxy/styrofoam blades.

4. Drive. Because of the low rotational speed of the turbine (under 200 rev/min), gearing must be used, to increase the rotational speed for the generator. It is confidently anticipated that this can be accomplished using toothed belts with only a moderate loss of power.

ELECTRICAL POWER TAKE-OFF CONSIDERATIONS

The electrical power output control of small (1-15 kW) single-machine isolated load, wind turbine systems is complicated by the variability of the energy source and the requirement to operate the wind turbine itself at variable speed if high efficiency is to be obtained. The design task is to ensure that throughout the normal operating range the wind turbine is loaded by the generator and load just enough to enable it to run at optimum rotational speed (i.e. the speed at which the power that is extracted from the wind is a maximum). This requirement can be met if the load torque on the wind turbine be arranged to vary as the square of the shaft rotational speed. A constant tip speed ratio can be achieved in a variety of ways, with varying degrees of sophistication.

CONTROL STRATEGIES FOR HIGH EFFICIENCY

(a) Wind Turbine Control:
  (i) variation of blade pitch.
  (ii) variation of gearbox transmission ratio.

(b) Electrical Generator Control
  (i) field control (applicable to d.c. and synchronous machine types).
  (ii) generator output control, e.g. d.c. chopper for d.c. outputs, thyristor bridge for a.c. outputs.

(c) Load Control
A control system employing a tachometer with threshold switching operations (operating in three to four steps) can be used to switch onto electrical load; this enables the square law characteristic of the wind turbine to be followed in a step-wise manner.

(d) Careful choice of wind turbine and generator such that the turbine is loaded by the generator and connected load just enough to enable it to run at optimum speed. In most cases even such systems as these are not entirely free of control means since special measures for starting and overspeed protection are often necessary.

CHOICE OF ELECTRICAL GENERATOR

The requirements of likely electrical loads may be grouped into two separate categories, (i) high quality energy, i.e. pure waveform at near constant frequency, (ii) low quality electrical energy for storage, i.e. batteries, storage heaters etc.

The choice of a particular type of generator will be determined by the load requirements and the generator/ancilliary equipment costs.

Studies have been carried out to examine the applicability of different generator types to meet the requirements of the loads specified in categories (i) and (ii) and the following information is available:

1) Load requirement - regulated voltage and constant frequency supply to an isolated grid under varying load and wind conditions.

   David and OOi (1979) have demonstrated the feasibility of operating an induction generator under such load conditions. The system also employs a synchronous generator which supplies the magnetisation current to the induction generator, which in turn, in addition to supplying the load power, feeds the synchronous generator with real power to replenish windage, friction and I^2R losses to sustain it at synchronous speed. The study is oriented towards the development of wind-power electrical systems in which multiple units of wind powered generators are connected to a common isolated grid. The system's main advantage is that a robust and reliable induction generator can be used to supply power whilst a lightly loaded synchronous generator is employed to attain a reference voltage and frequency. In view of the lack of work carried out to determine the stability and dynamic characteristics of such systems, a diesel back up would be considered necessary.

2) Electrical energy storage - Wind energy systems with battery storage and diesel back up for isolated communities has been looked at by Slack, Sexon, Collins, Dunn, Lipman and Musgrove (1979). The authors conclude that small consumers, remote from a grid system in windy areas, may well have their electricity met at lower cost by a wind energy system. They investigate the effect of wind turbine size/battery capacity mix, under various economic conditions, using a year's hourly mean wind speed data for a selected site.

The factors relating to the choice of electrical generating systems for small wind turbines has been examined by Bolton and Nicodemou (1979). These authors conclude that if minimum capital cost per rated kW is the primary criteria for equipment selection, low cost generators may be chosen, even though they may convert energy relatively inefficiently. Where minimum cost per kWh is the primary criterion, the use of generators of the highest efficiency and reliability is normally justified and permanent magnet generators have a useful role to play in small wind plants.

Bolton and Nicodemou (1979) have examined the operation of self-excited generators for wind turbine application. They deal with the operation of self-controlled wound-field generators for small-scale wind-power application in which the output is connected to a constant-resistance load and the field is connected in plain shunt or series. It is shown that, with this configuration, a fairly good natural match can be obtained between the power/speed characteristics of the generator and the wind turbine.

FUTURE ACTION

It is the general intention of the research programme to investigate the performance of vertical axis wind turbine configurations of low aspect ratio and high solidity. Blade stressing will also be completed.

In parallel with this will be an investigation of the symbiotic interaction of an induction generator and synchronous generator with a static thyristor Scherbius drive circuit connected into the wound rotor circuit of the induction machine.

It is hoped that within three years a satisfactory design configuration will be achieved. Principal problems lie in the field of detail design rather than in any essentially innovative direction.

REFERENCES

Bannister, W S (1979). Aerodynamic studies of a straight-bladed vertical-axis wind turbine. Proc. 1st British Wind Energy Association Wind Energy Workshop. (April 1979).

Bolton, H R and V C Nicodemou (1979). Proc. 1st British Wind Energy Association. Wind Energy Workshop. (April 1979).

Bolton, H R and V C Nicodemou (1979). Proc. IEE, Vol. 126, No 9, September 1979.

David, R A and B T Ooi (1979). Induction-generator/synchronous-condenser system for wind-turbine power. Proc. IEE, Vol. 126, No 1, January 1979.

Healy J V (1978). The influence of blade thickness on the output of vertical axis wind turbines. Wind Engineering Vol. 2, No 1. 1978.

Mays, P I and P J Musgrove (1980). Performance of the variable geometry vertical axis wind turbine at high and low solidities. Proc. 2nd British Wind Energy Association Wind Energy Workshop (April 1980).

Miller, D (1980). Storing wind energy as a liquid fuel. Proc. 2nd British Wind Energy Association Wind Energy Workshop (April 1980).

Musgrove, P J., and I D Mays (1978). Development of the variable geometry vertical axis windmill. Second International Symposium on Wind Energy Systems. Paper E4 organised by Brit. Hydromech. Res. Association, Cranfield (October 3-6th 1978).

Slack, G., B Sexon, R Collins, P Dunn, U Lipman and P Musgrove (1979). Proc. 1st British Wind Energy Association Wind Energy Workshop (April 1979).

Aspects of a Wind Energy Conversion System

B. Sexon, G. Slack, P. Musgrove, N. Lipman and P. Dunn

Department of Engineering, University of Reading, UK

Introduction

The high cost of independent electricity production for rural, isolated communities has meant that increasing interest is being shown in the possible use of renewable energy sources for this application. The energy system being considered at Reading University uses a wind turbine as its primary power source, Fig. 1. This is backed up by a short term battery storage facility and a diesel generator. The batteries store any excess energy from the turbine, delivering it to the load if the turbine is unable to produce sufficient power at any instant. The diesel generator ensures the continuation of supply during any lengthy period of calm which may occur. The size range of community being considered would have a power capacity of from 10kW to 50kW.

The system is to be designed using equipment which is commercially available now. It will provide the electrical requirement for a community rather than energy used for cooking or space heat. For this reason batteries were chosen as the storage medium, although alternatives were considered, as described below.

Storage

The medium used for storing excess energy from the windturbine is of particular importance in a small system. Efficient use of a storage medium will significantly influence the percentage of the turbine's output which can be utilised. There is a wide range of methods for storing energy, many still at the development stage. Flywheels and compressed air are among this category. The use of flywheels is a very long term proposition and if they are introduced it will probably be only for small amounts of energy storage. Compressed air storage is likely to be economically viable only when used on a very large scale.

Another storage medium, hydrogen, has not yet been used in conjunction with wind turbines, but since the means of generating hydrogen by electrolysis are well established its possible use in the near future should not be ignored.

Hydrogen

Hydrogen and oxygen gases can be generated by the electrolysis of water. An electric current passed through a solution of caustic soda (NaOH) or caustic potash (KOH) in distilled water will cause the evolution of hydrogen at one electrode and oxygen at the other. It is not necessary for the current to be absolutely uniform; the electrolyser will operate through a range of from 30% to over 125% of its rated power requirement. The energy efficiency at 30% load is 71% and at 125% load is 58% since the energy demand per m^3 hydrogen produced increases with the percentage of rated power being delivered to the electrolyser.

A cost analysis of a windturbine/electrolyser system showed that for a large system using an electrolyser with a rater power demand of 250kW, it is possible to produce hydrogen at 16p/m^3 (suffix n refers to a temperature of 0^oC and 1 atmosphere pressure). This would involve a capital outlay of approximately £400,000 and would produce 290,000m^3 of hydrogen per annum. The system includes a wind turbine, electrolyser and water purifier, but no compression or storage facility. A small system using an electrolyser with a rated power demand of 10kW can produce hydrogen at a cost of 26 p/m^3 for a capital outlay of £32,000. This system will produce 23500m^3 of hydrogen per annum Fig. 2. shows a graph of cost per m^3 of hydrogen for varying rated to mean wind speed ratios for the wind turbine.

The hydrogen is produced at 14 bar pressure, high enough for low pressure distribution in a gas system for cooking and space heat. If the hydrogen were to be stored in cylinders a compressor would have to be included, increasing the cost considerably.

Hydrogen is a very versatile storage medium; it can be used directly for cooking and space heating or used in an internal combustion engine for transport or to generate electricity. Although the use of hydrogen for all these applications has been proven, the adapted equipment necessary for use with hydrogen is not generally available. Therefore, hydrogen as a storage medium is a possibility for future development, especially as other fuels become more scarce. However, as stated previously, the system being designed is for current use so it was decided that hydrogen is not a suitable storage medium for this application. The only other mediums for storing wind energy which are established are water and batteries. Both of these are used for the direct generation of electricity and would therefore be suitable for the system being considered here.

Hydro storage

The generation of electricity using water power has been widely exploited for many years. When used together, a wind turbine and hydroscheme can operate in one of two modes. The wind turbine can be used to supplement the hydroscheme, so that a larger proportion of water is retained in the reservoir. This is a simple one way system using the hydroscheme and wind turbine together. Alternatively, a more complex, two way mode can be used in which the wind turbine is the primary source, its excess power being used to pump water from a low to a high reservoir. The hydroscheme is then only used to make up any power deficiency during calm periods.

The main disadvantage of using water as the storage medium for an electrcity generating scheme is the limitation on suitable sites due to the required head of water. Although it can be used in many isolated communites, hydro-storage is not particularly suitable for general applications.

Batteries

Batteries represent the simplest method of storing a variable input of electricity which occurs with a wind turbine. They can be used throughout the world, regardless of the terrain and are at present being used with wind turbines in a variety of locations. Research is being done into alternative battery designs but lead acid and nickel cadmium are the only types widely available for storage applications. Nickel cadmium batteries are very robust and are able to endure high charge/discharge rates without adverse affects, but their high cost makes them unattractive when compared to lead acid, Fig. 3.

Of the three main types of lead acid battery on the market, only the tubular sleeve and the Plante type are suitable for use with wind turbines. The pasted-plate battery, used for car starting systems, is totally unsuitable for any sort of regular deep cycling, as the plates tend to disintegrate if subjected to repeated charge/discharge cycles. This is caused by increased pressure between the plates due to a build up of sulphates during discharging.

Tubular sleeve batteries, for electric vehicles, were specifically designed for deep cycling, where as Plante are used for standby applications. Both types have been successfully used with wind turbines, although greater cycle life of tubular sleeve batteries, 2000 cycles makes them preferable to the Plante, which can endure cycles 700-1000 (Ref. 1).

The main factors affecting the life of a lead acid battery are the depth of discharge and the rate of recharge. During regular deep cycling of battery cells, the maximum discharge should be to 80% of the nominal capacity of the battery. Persistant completed discharging can lead to early deterioration. Continued high recharge voltages will also shorten the battery life and the maximum should be 2.4V per cell for a lead acid battery. This will restore 75% of nominal capacity with 6 hrs. to a completely discharged battery. Fig. 4 gives an indication of the effect of the depth of discharge on battery life from different sources (Ref. 2 and 3).

The cost of batteries could prove a significant proportion of the total system cost. As shown in Fig. 3 lead acid batteries will cost from £110-315 per kW supplied dependent on the size of the battery store. If the total energy delivered during the life of a battery is considered, then the minimum cost is 11p/kWh supplied. Fig. 5 (Ref. 4). This cost will depend on battery life and on the treatment to which the battery is subjected. It is obvious that the amount of energy being put through the battery store should be kept to a minimum. This fact and the limitations on the rate of recharge will affect the amount of energy from the turbine which can be stored. It may be necessary to dump some of the output from the turbine in order to increase the useful life of the battery store.

Computer Model

As part of the study into the wind energy system, a computer model was set up representing the system shown in Fig. 1. The operating philosophy of the model is as follows. - All electricity from the wind turbine is primarily directed towards the load, any excess being stored in a battery bank. If the wind turbine is unable to supply the total load, the battery bank is discharged into the system. If a state is reached where the combined output from the batteries and the wind turbine is unable to meet the load requirement, a diesel generator is brought into the system. The diesel generator always runs at its optimum power level, any excess being fed back into the batteries.

Fig. 6. shows the cost of electricity for a 'typical' small community using this model. The variable demand on the system Fig. 7, had a mean value of 8kW and a load factor of 0.3. Using the computer model it was shown that the optimum size of wind turbine swept area is $78m^2$. For this size of turbine, and using a battery store large enough to supply the mean load for 10 hrs, the cost of electricity is 13.6p/kWh. This represents a capital outlay of £47,000, discounted over 10 years for the batteries and 20 years for the rest of the equipment.

Fig. 8 shows the energy distribution through the system. 38% of the total load is supplied by the wind turbine direct, 35% from the diesel generator and 27% from both via the battery store. This diagram shows the energy losses incurred during rectification and inversion to and from the battery and also the internal losses in the batteries. Provision for the dissipation of the excess power from the wind turbine is assumed.

The minimum value of 13.6 p/kWh was achieved using one set of system charateristics. This is 10 hours worth of battery storage and a rated to mean wind speed ratio for the turbine of 1.5. By reducing the ratio of rated to mean wind speed and increasing the wind turbine size, so that the rated power output is unaffected, it is possible to increase the percentage of the year during which rated output is achieved. This, in turn will decrease the battery storage requirement. A detailed analysis has to be done to find the optimum values for the rated to mean wind speed ratio and the size of battery store. By doing this it should be possible to reduce the cost of electricity from the system to below the present minimum. A detailed description of the computer model can be found in a report produced for the United Nations (Ref. 4).

Future Work

Although the performance of a wind turbine/diesel system can be determined and analysed to a certain extent using a computer model, this does have its limitations. It is therefore hoped to perform some tests on a system using a simulated wind turbine with varying battery storage and diesel generator configurations.

The characteristics of a wind turbine are well defined and the electrical output from a turbine can be simulated accurately. This has the advantage that different wind regimes and turbine characteristics can be put into the system. The experiments would be independent of the vagaries of the wind on site and tests could be performed as and when required. It is hoped that the experiments would determine the operating characteristics of battery storage when subjected to a varying and erratic charging regime. All information so far available concerning battery performance deals with constant recharge conditions only. By varying the limitation on rate of charge to the batteries it should be possible to determine the optimum level in terms of energy storage during life.

Life test on batteries for this application will prove more difficult. A large number of batteries would have to be tested to failure under varying conditions before any justifiable conclusions could be drawn. Comparison tests between Nicad and lead acid batteries may also be performed to decide whether there is any justification for spending the extra money to obtain the higher performance nickel cadmium batteries for a wind turbine application. Indeed, whether it is necessary to use tubular sleeve batteries, or if the Plante type are sufficiently robust.

Refs.

(1) Selection of Batteries for Stationary power applications. — G.M. Bulman and K. Lay.

(2) Batteries and Energy Storage — C.L. Martell.

(3) Private Communication — Chloride Alcad.

(4) Wind Turbines/Diesel Electricity Generation for Small remote communities. — B. Sexon, G. Slack, P. Dunn N. Lipman, P. Musgrove. Reading University.

—— // ——

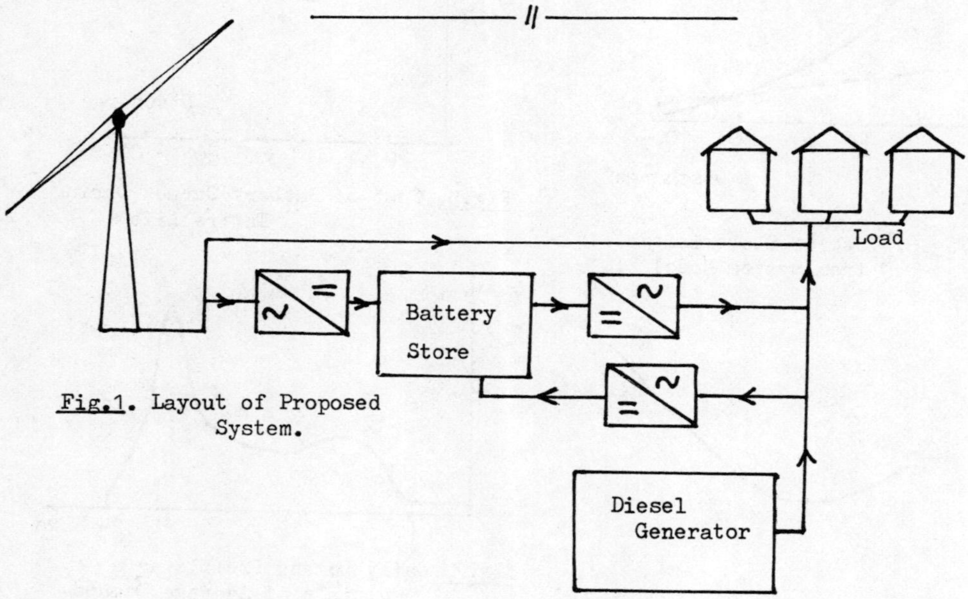

Fig.1. Layout of Proposed System.

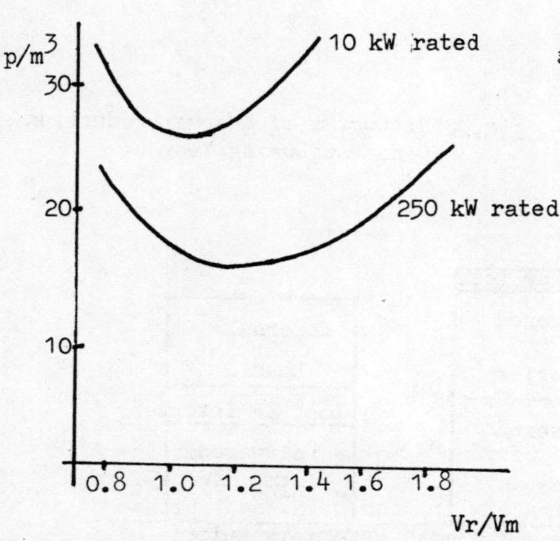

Fig.2. Cost of Hydrogen Production pence per m^3

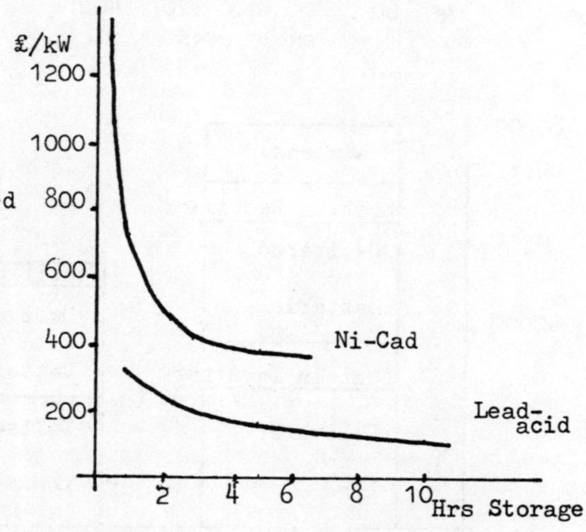

Fig.3. Battery Storage Costs £/kW of power supplied

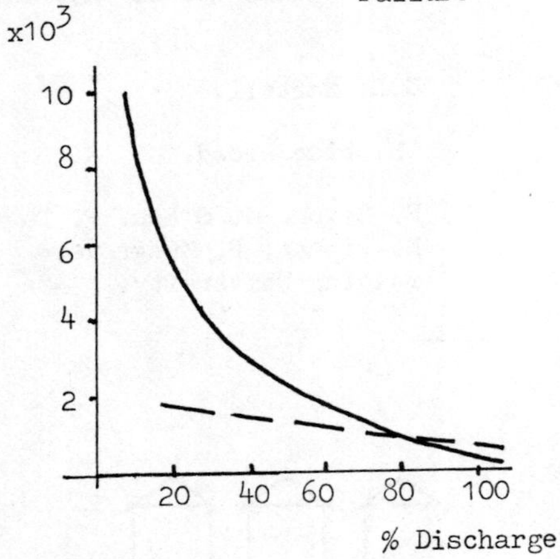

Fig.4. Number of Cycles to Failure

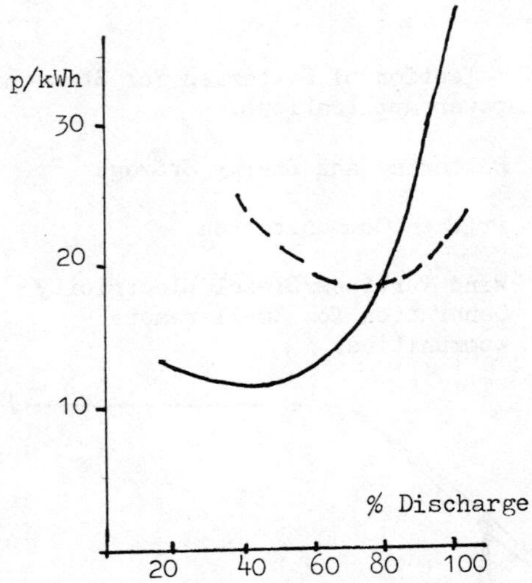

Fig.5. Cost of Battery Output During Entire Life

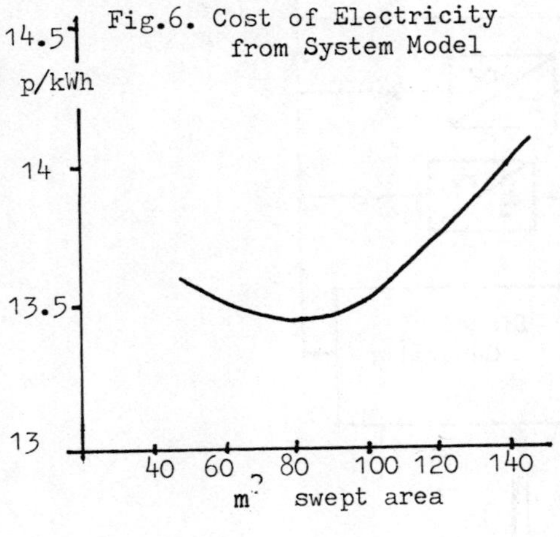

Fig.6. Cost of Electricity from System Model

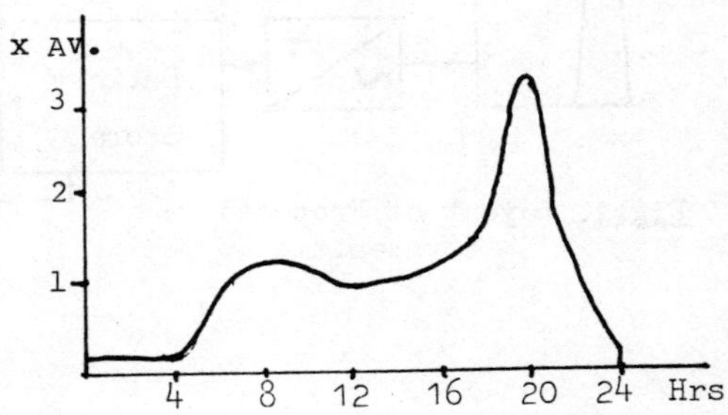

Fig.7. Daily Demand Profile as a Multiple of Average Demand

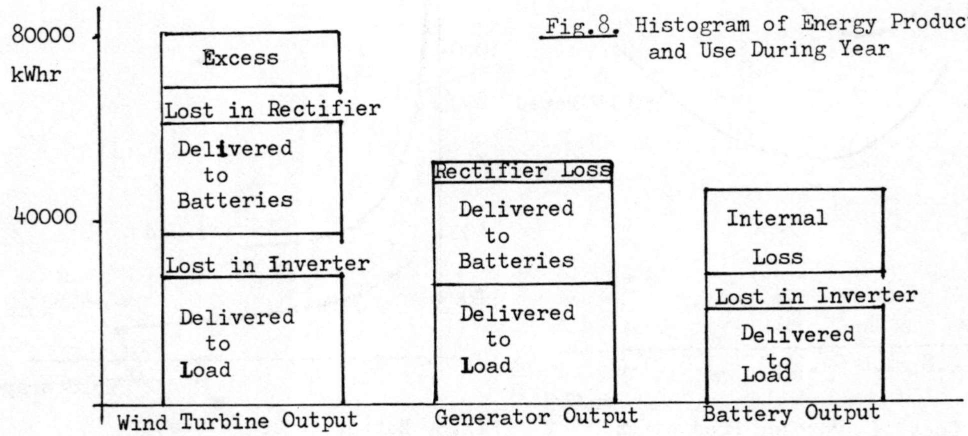

Fig.8. Histogram of Energy Production and Use During Year

Commercial Wind Turbine Generators and their Possible Integration into Weak Networks

L. J. Ballard

ERA Technology Limited

ABSTRACT

Increasing fuel costs have made alternative sources of energy economically viable in a wider range of applications. This is particularly the case for some island communities where the delivered cost of fuel is high.

Medium-scale Wind Turbine Generators (WTGs) for operation in parallel with an electrical supply have recently been put into production by several companies. Details of selected units are presented. Operational history of previous commercial WTGs is discussed and from consideration of some aspects of interaction of WTGs with a power system, features of a WTG which are preferred for reliability and stable network operation are suggested.

KEYWORDS

Alternative energy, wind power, electricity generation, power systems.

INTRODUCTION

Increases in fossil fuel prices have meant that alternative sources of energy such as wind power are becoming economic in a wider range of applications. This is particularly the case for some island communities where direct electrical connection to a large electrical supply grid is impractical and where the delivered cost of fuel is high.

Several companies, realising the potential for economic generation of electricity from the wind, have begun to manufacture WTGs for operation in parallel with an electrical supply network. As the wind is by nature a variable source the integration of the wind-produced electricity into the electrical network poses problems of power control and adequate matching of power supply and demand. This is particularly critical with island communities where electrical systems tend to have low installed capacity, a wide variation between minimum and maximum demand, and high impedances at many points on the system.

This paper outlines WTGs which are commercially available and suitable for network connection. Some of the problems and possible solutions to the integration of wind power into a small power system are highlighted.

A review of presently available commercial medium scale WTGs by Buss and Hardy (1980) has shown that there are 11 leading manufacturers who can now offer some 15 machines suitable for network connection (Table 1). 'Medium-scale' is broadly categorised as machines having a turbine diameter of 5 m to 15 m corresponding to a power range of approximately 5 to 55 kW.

The commercial development of these machines is proceeding rapidly (particularly in Denmark, W. Germany and USA) and the list shown in Table 1 will change and generally expand in the next few years. The list is however generally representative of the types, sizes and ratings of machines that will continue to emerge in the near future. The details are broadly accurate for the beginning of 1980.

The features of a WTG are important within the context of this paper from the separate aspects of machine integrity and performance when operating on a network. Some of the machine features which provide for maximum integrity and reliability are however not necessarily those which would be chosen for most satisfactory network operations. The preferences are highlighted in the following sections:

MACHINE INTEGRITY

Experiments, R & D and commercialisation with medium scale WTGs have been conducted on an organised and well documented basis as far back as 1890, and there is a considerable (if dispersed) body of information and experience upon which to draw with regard to the performance and behaviour of different machine types.

A comprehensive discussion of the results obtained from these different machine types with reference to the principal design and construction features used is well beyond the scope of this paper, but it is worthwhile observing a few trends.

The major component in any discussion of integrity must be the WTG itself. The commonly stated design options are number of blades, fixed/variable pitch, fixed/variable speed, horizontal/vertical axis type, blade materials and construction alternatives, operation upwind or downwind of the tower, and various options in the means of orientation to follow the wind direction. Whilst these all have a bearing on reliability, the most crucial features seem to be whether or not to use pitch variation in the turbine, and how many blades the turbine should have. It is assumed that insufficient experience is available with vertical axis turbines at this stage to adequately comment on the associated integrity issues.

Also important in consideration of reliability, although perhaps not quite so obvious, is the choice and type of generator. In the past, dc generators have been extensively used, but the main choice in modern WTGs is between ac synchronous and induction types.

The principal manufacturing and operational experience with medium-scale WTGs has been substantially within four types or groups of machine. The earliest were produced by Lykkegaard (Denmark) in the period up to 1914. These machines (between 7 m and 18 m diameter and rated 3 to 30 kW) used four 'traditional' windmill sails with automatic slat regulation and incorporated dc generators. They have little relevance to modern technology or needs but are worthy of mention because several hundred were produced, some being in service until the 1940s.

More relevant are the Danish successors to these machines, commonly known as the Aermotors, produced by F L Smidth in the 1940s. The main features of the two machines produced are given in Table 2. The Aermotors proved to be reliable and robust, many of them are still standing (albeit in varying states of repair)

TABLE 1 COMMERCIAL MACHINES FOR NETWORK CONNECTION

MANUFACTURER	COUNTRY	MODELS	TYPE	DIAMETER (M)	FIXED OR VARIABLE PITCH	NUMBER OF BLADES	RATED POWER (KW)	RATED WIND SPEED (M/S)	GENERATOR DRIVE	GENERATOR TYPE	PRODUCTION STARTED	NUMBER PRODUCED
DAF INDAL	CANADA	37X55	VA	11.2	FIXED	2	56	14.8	GEARBOX AND BELT	INDUCTION	1978	3
DYNERGY	USA	5M	VA	4.6	FIXED	3	8	10.7	DIRECT	PERMANENT MAGNET AC	1977	~10
KONGSTED MOLLEN	DENMARK		HA	10.0	FIXED	3	22	5.0	GEARBOX	INDUCTION	1979	5
ENERGY DEV CO	USA	440 445	HA HA	11.0 12.2	FIXED PRE-SETABLE	4 4	40 45	11.2 11.2	GEARBOX GEARBOX	INDUCTION INDUCTION	1976 1976	18
MILLVILLE	USA	10-3-IND	HA	10.0	FIXED	3	10	11.0	GEARBOX	INDUCTION	1977	~35
POULSEN	DENMARK	30KW	INCL-INED	13.0	FIXED	2	5.5/30	11.0	GEARBOX	2 X INDUCTION	1978	4
VESTAS	DENMARK	10 30/5	HA	10.0	FIXED	3	5/30	11.0	GEARBOX	2 X INDUCTION	1980	3
VINDKRAFT	DENMARK	55 A 15	HA	8.4	VARIABLE	3	15	9.5	GEARBOX	INDUCTION	1977	5
WIND ENG CORP	USA	WINGEN 25	HA	12.2	VARIABLE	3	25	11.2	GEARBOX	INDUCTION	1980	5
WIND MATIC	DENMARK	12/5-22 18/5-30 18/6-45 18/7-55	HA HA HA HA	10.0 10.0 12.0 14.0	FIXED FIXED FIXED FIXED	3 3 3 3	22 30 45 55	12.0 12.0 12.0 12.0	GEARBOX GEARBOX GEARBOX GEARBOX	INDUCTION INDUCTION INDUCTION INDUCTION	1976 1976 1976 1976	~60
WIND POWER SYSTEMS	USA	10-8-1G-3P-60	HA	10.0	FIXED	3	18	8.0	GEARBOX	INDUCTION		~ 4

VA - VERTICAL AXIS HA - HORIZONTAL AXIS

throughout Denmark; the infrastructure of the machines is sufficiently sound that refurbishment is now being considered for some units. It is likely that the success of these machines (their development culminating in the famous 200 kW, 24m Gedser machine) has been a strong influence behind the dominant tendency in modern Danish WTGs towards three bladed fixed pitch turbines.

TABLE 2 FEATURES OF THE AERMOTOR, D-12, D-18 AND ALLGAIER MACHINES

Model	AERMOTOR		D-12	D-18	ALLGAIER
Electrical Rating (kW)	50	70	8.6	22	6 to 8
Number of Blades	2	3	3	3	3
Blade Pitch	Fixed, with regulation by auxiliary control surfaces		Fixed, with regulation by auxiliary flaps + rotating tips		Variable controlled at the hub
Turbine Diameter (m)	17.5	24.0	12.0	18.0	10.0
Generator	Differentially Compounded DC		Synchronous		DC, or dual output (2 generators) DC + Synchronous AC

Lesser publicised but equally important in operational experience are the D-12 and D-18 machines produced in the USSR during the 1940s. Table 2 gives outline data.

It is probable that these machines were designed for connection to high impedance points of the supply network and this could explain the choice of synchronous generator. Interestingly, however, later versions of the D-18 incorporated an induction generator. The success of these machines is difficult to judge but adoption of similar principles in the 12 m diameter 'Sokol' USSR machine in the early 1970s could be taken as a vindication of the basic design.

The fourth production medium-scale WTG of note was the 'Allgaier' machine made in West Germany in the 1950s. Data are shown in Table 2. The Allgaier machine was a sophisticated unit, incorporating a centrifugally initiated hydraulic pitch control at the hub. Whilst apparently successful in its production (over 100 units were sold), there were operational problems, especially with contamination of the hydraulics (for example by sand and dust) leading to eventual failure. The machine procured and used by ERA at Aviemore (Walker 1960) eventually suffered a generator failure.

Apart from the larger prototype units built in Western Europe and USA during the 1940s and 1950s, which are well documented but of less relevance to this paper, the medium-scale machines which have been reviewed form the backbone of any historical experience assessment.

The principal conclusion to be drawn from this early experience is that for reliability it is most important to use the simplest and most rugged possible turbine arrangement. Fixed pitch blades are preferred, power regulation or overspeed limiting being achieved by small ancillary control surfaces such as flaps or rotating tips. Most turbines in this class have incorporated three blades and there is not much basis for comparing the performance of two and three bladed units. However, notable success of a number of three bladed machines in the past may be due to the relative ease with which a three bladed fixed pitch turbine can be braced and stiffened.

The selection of generator type is less easy to support from an examination of past experience. DC machines, which have been extensively used, are no longer suitable for network connection (unless an expensive inverter system is incorporated) and the brush maintenance problem is now generally felt to be unacceptable. The choice lies between ac synchronous and induction generators and both have been used in this application. The synchronous machine is of more complicated construction and requires associated excitation control equipment; this directly implies a lower degree of reliability than the induction generator, which is a simple, robust machine needing no voltage control equipment and identical physically to the ubiquitous industrial induction motor. It seems likely also that since the induction machine operates with a small 'slip' in speed which varies with power delivered, that some of the transient wind energy incident on the turbine can be absorbed and stored as kinetic energy as the speed rises during some short term increase in wind speed. The turbine inertia represents a useful 'buffer' in this respect, and the absorption of energy can reduce the level of transient torques in the transmission and the rapid fluctuation of electrical power output. This possibility was set out at an early stage by Kaspar (1945). The synchronous generator by contrast can only accommodate very small transitory speed change, corresponding to a shift in load angle, and relatively high transient torques and output power fluctuations are possible. This reinforces the conclusion that for reliability an induction generator is preferred.

Overall then the preference from reliability considerations is for a WTG with fixed pitch blades and an induction generator. Although the present commercial producers of medium-scale WTGs are all new to this technology and have probably not unearthed much of the evidence of the past and the basis for their design decisions, it is not surprising to find that 12 out of the 15 machines listed in Table 1 are fixed pitch with induction generators and of these 12, 9 incorporate three-bladed turbines!

SYSTEM CONSIDERATIONS

The WTGs cannot be considered alone since they will inevitably affect the electrical system to which they are connected. Two important aspects which will affect the operation of a system containing WTGs are power and voltage control.

On an electrical system containing only conventional generating plant a change in demand results in an increase or decrease in the amount of fuel supplied (normally through the automatic action of governors and throttles or steam valves) to the prime mover so that the power output is controlled to match the new load demand. For a system including WTGs the situation is more complex since one of the power sources (in the wind) is variable, uncontrolled and unpredictable on a short term basis. A WTG can be envisaged in this sense as a negative load on the system.

Variations in power output of the WTGs will ideally be matched by changes in the power output of the conventional generating equipment running in parallel, so that the total power supply can be continually matched to the load. This may not always be possible particularly for power output changes over short time periods (tenths of seconds) less than the response time of the conventional generating equipment. WTG output variations over longer periods will be accommodated through power control on the conventional generators. In order to provide the required capacity, should the wind speed fall, the conventional generators running in parallel will be only partly loaded during wind generation periods.

Power control is important in a power system since excessive power loadings can lead to system instability, the shedding of generation equipment and ultimately the loss of supply. It is a factor which may limit the capacity of wind driven generation that a system can tolerate, and may influence the type of WTG that can be used, or whether any ancillary control or storage measures are necessary.

A element of power control is introduced at the wind turbine if pitch variation or some other means of aerodynamic torque control is used. This feature is advantageous for the system since it provides the flexibility to regulate the power output of the WTG in a way that is analogous to the governor on a conventional prime mover (with the exception that for a WTG the maximum power available varies continuously as the wind speed changes). A fixed pitch turbine offers very little power control since the stalling characteristic at fixed rotational speed will place a limit on the power output and some form of airbrake is incorporated to prevent overspeeding.

Another means of achieving power control is to vary the demand on the system through load switching. In such a scheme loads are connected to the system on a priority basis in merit order depending on the wind power available. Supplies to high priority loads such as lighting, TV etc. would be connected continuously, whereas supplies for low priority loads such as refrigerators and water heaters would be interruptible. Centralised load control is being introduced on some supply networks in conjunction with water heating, and proprietary equipment is now available. This is similar in principle to the operation of so called load management controllers being introduced into industry. These switch industrial loads so that the load-duration curve of a plant or factory is levelled resulting in a reduced maximum demand.

The related issue of voltage control is important in relation to the quality of supply to consumers. In a conventional power system voltage stability is maintained by synchronous generators which incorporate automatic voltage regulators (AVRs). An AVR senses reduction in system voltage and increases the excitation or field current to the generator until the prescribed voltage is re-established. Large systems have power generation at numerous points throughout the system and because the impedance of these systems tends to be low the voltage stability achieved is good. Smaller systems tend to have fewer, more centralised generation plants and because of the higher impedances associated with these systems the voltage stability at the point of consumer supply can be more of a problem to maintain within acceptable limits. If WTGs are to be introduced into such a 'weak' system it would be advantageous if they could improve or at least maintain the voltage stability of the system.

A WTG incorporating a synchronous generator and AVR can reinforce the voltage stability of the power system. The induction generator as used in the majority of medium-scale commercial WTGs, obtains its excitation directly from the network. Independent voltage control is not practicable and the induction machine, without auxiliary control measures, will not improve and may reduce the voltage stability of the system.

The reactive power (var) requirement for induction generator is typically in the range 40-60% of the rated power of the machine. This large wattless current not only introduces ohmic losses in the power system but will also create difficulties in the voltage regulation of the system. This can be overcome to some extent through the use of local reactive compensation using for instance switched capacitor units.

The fuel saving benefits of network connected WTGs are partly eroded by the poor part load efficiency of diesel generator sets, especially those of lower rating. It would be advantageous to shut down the conventional equipment at times of high wind power availability, but in this case some alternative means of governing the power system must be provided. This can be achieved by the introduction of an energy storage facility, with appropriate output power control. However, the economics of conventional storage systems such as electrolytic cells are very discouraging for all but very short term storage needs. Other storage techniques commonly quoted, such as compressed air, pumped water, hydrogen, flywheels and thermal phase change systems are generally speaking either not well enough developed, site dependant, uneconomic, inefficient, over-demanding on maintenance and supervision, or inappropriate for small scale operation.

Overall then, the preference from the system considerations is for WTGs which provide both power and voltage control. This implies the selection of a machine having variable pitch blades and a synchronous generator.

CONCLUSIONS

The current range of commercial, medium-scale machines available for network connection have a short production history and are relatively unproven. Previous WTGs in this class although no longer manufactured, and often including features such as dc generators, inappropriate for modern applications, do however provide useful operating experience on the design and reliability of the turbine.

Two options for turbine design have been highlighted, these being fixed pitch, (with overspeed limiting and power regulation through the use of auxiliary control surfaces), and variable pitch blades which provide power and overspeed control through pitch variation. From the operational history of the commercial machines considered, the fixed pitch turbine constitutes the most reliable, proven and robust design. The incorporation into a fixed pitch turbine of auxiliary moving surfaces would not seem to detract from the integrity of the turbine and does offer the possibility of power regulation without the complexity of complete pitch variation using articulation at the turbine hub.

The facility of power control in a wind turbine is important, so that the stability of the connected power system is maintained.

Also important is the need for voltage and reactive power control particularly where the system is of high impedance. The related choice of generator centres around synchronous and induction machines. The induction machine is more robust and simpler but does not provide voltage and reactive power control. The synchronous machine is more complex but readily allows independant voltage control. The need for voltage control in 'weak' networks probably dominates the reliability considerations and favours the selection of a synchronous generator.

Trends in present commercial machines seem to be broadly categorised as (1) fixed pitch with overspeed control with an induction generator, and (2) variable pitch with a synchronous generator.

The most attractive option at this stage however appears to be a fixed pitch turbine with some form of power regulation (such as brake flaps) combined with a synchronous generator which allows power, voltage and reactive power control.

REFERENCES

Buss, B and Hardy, W E (1980) 'A review of medium-scale wind turbine generators.' BWEA Second Annual Workshop, Cranfield April 1980

Walker, J G (1960) 'The automatic operation of a medium-sized wind driven generator running in isolation'. ERA report C3T 122, 1960

Kaspar, F (1945) 'The asynchronous generator and its use in wind power stations' Elektrotech. Obs. Vol 34, No. 4, pp 60-63 and No. 5, pp 81-89 (In Czech) (ERA translation Trans/IB 1051)

Large and Small Electric Wind Turbines for Isolated Applications

Kenneth Bosley

American Refining Group Inc., York Road, Sparks, Maryland 21152, USA

ABSTRACT

Practical planning consideration of electric wind turbines at isolated facilities are reviewed. Topics discussed include concerns of siting, integration with other generation equipment; power quality, shipping, installation maintenance and operational advantages. Today, isolated installations provide the most cost effective utilisation of wind turbines. In the future, isolated areas will provide the resource of land agrea with high winds for large scale power production which will be transmitted to major population centres.

KEYWORDS

Wind; Generation; Isolated.

INTRODUCTION

A main consideration of where man constructs his facilities is finding places where there are low winds. Thus isolated areas tend to have high wind. Isolated sites have three general scenarios for acquiring power. The first is that they may install transmission lines from a major grid. The second is to transport fuel in and operate an independent grid. The third is utilising locally available resources for power production. Any combination of these three scenarios may be appropriate. In the past, isolated diesel grids were interconnected with major grids to acquire "inexpensive power" from the major grids. With the current progress of power production from locally available resources in isolated areas, the isolated grids are either staying independent or they are becoming net power producers to the major grids. As power from the major grids becomes more and more expensive, the power from the rural areas is more likely to be cheaper than that from the major grids. A typical situation demonstrating this trend may be an island with an independent diesel grid. Then suppose wind turbines are installed to save fuel and decrease the maintenance on the diesel generators because the island has sufficient available land and winds greater than that of the adjacent main land.

The major grid on the mainland justifiably needs power from wind turbines but lacks the necessary high winds and inexpensive land. A submarine cable is thus laid between the mainland and the island. Many more wind turbines are installed on the island to provide net power to the mainland in large quantities. Two commercially available wind turbines are shown in figures 1 and 2.

Since the power in the wind increases cubicly, acquiring a site with high wind velocity is the only way to have a high level of generation. Daily and seasonal wind speed characteristics need to be considered with load requirements. The least expensive power quality requirements also need to be assessed. Installation and maintainability requirements of course have an extreme impact on the feasibility of wind turbines.

WINDSPEED AND LOAD CHARACTERISTICS.

When siting a wind turbine many things must be considered and traded off. The wind turbine must be well exposed. This requirement dictates usually that the wind turbine must be highly visable. The surrounding area of the wind turbine tower must be cleared of trees and other obstructions to assure high wind flow with minimum turbulence. Ice that forms on wind turbines will be thrown a short distance.

Typically tropical areas with trade winds have moderate winds very consistantly. Areas in the higher latitudes typically have higher average winds but of more variability. During periods of low wind, supplemental power must be provided by storage or other generation equipment. Typically, for loads greater than that of a single residence, internal combustion engines provide less expensive power than batteries. Short term storage with batteries for larger systems may be justified in some instances. Islands may have some wood for gasifiers and/or some water flow for hydro-electric turbines but not enough to meet the total power requirement needs. Wind turbines would have an important role in this situation and would permit the occasional use of the wood and pondage of the water flow to be only used during periods of low wind and/or high demand. When the wind turbines are producing more power than the load demands, the excess may be stored as heat or pumped hydro. It is often overlooked in load prioritization systems that a diesel generator may use three times more oil producing a unit of electrical resistive heat than an oil burner requires to produce the same heat. A community facility of sufficient size would likely be able to justify a control system to utilize resistance heating whenever the non-heating electrical demand is lower than the production from the wind turbines.

POWER QUALITY OPTIONS

There are several considerations concerning power quality. Voltage and frequency have tolerance requirements dependent upon the equipment that is being powered. Transmission line losses are increased with the use of AC instead of DC, and low voltage instead of high voltage. Frequency may be controlled to produce power at the highest quality tolerances even when the wind turbine is the dominant generator on the grid. Synchronous inverters will maintain frequency as long as a diesel generator of three times the capacity is at least idling on line. A static inverter requires that its power be drawn from a battery bank, whereas a synchronous inverter may draw its power directly from the wind turbines generator. A static inverter will maintain frequency and voltage tolerance even if it is the only power source on the line. Static inverters are more expensive than synchronous inverters, and in addition the cost and efficiency of the static inverter system includes a battery bank. In a grid with resistance heating and inverters, it would be advantageous to investigate the trade-offs of separate wiring to the resistance loads that would by-pass the inverter. Likewise, certain motors may be run with variable frequency AC. The mechanical drive shaft of the motor would vary with the frequency but this may be acceptable to operate certain equipment for certain applications.

INSTALLATION AND MAINTENANCE.

Isolated sites tend to have insufficient roads and local facilities. Therefore self-propelled cranes cannot be brought to the site. Various installation techniques involving engine powered winches and maritime rigging techniques have been devised. Wind turbines up to 38,000 kg have been installed this way. A large amount of concrete for foundations or towers is not practical for isolated sites. Hauling the vast quantities of material and the logistics of mixing it are beyond the capabilities of the infrastructure. Large wind turbines at isolated sites may be installed with grillage type of foundations which only require a small base of concrete for levelling purposes. The metal grillage is then fabricated and earth is compacted over it. At some remote parts of the world there are high altitude facilities which are serviced only by light duty helicopters that can carry up to only several hundred pounds at a time. This requires that certain assemblies must be able to be disassembled into smaller components and re-assembled under site conditions. For larger aircraft, there are standard shipping container sizes that wind turbine assemblies should be designed to fit in. Once the machine is installed, provision must be made for its maintenance. This may require only that fittings must be greased every year. Certain spare parts may be stocked locally or regionally. Two wind turbines instead of one of twice the size for very isolated locations may be in order. Another option is to have one large wind turbine and have parallel redundant electronics installed.

OPERATIONAL ADVANTAGES OF WIND TURBINES.

A wind turbine interconnected with a diesel will decrease the operating time of the diesel and thus overhauls will be less frequent. Along with decreasing fuel usage, time between refuellings would be increased. This is an important consideration when the diesel fuel is flown in. Another advantage of the wind turbine has become evident for unattended sites such as telecommunications facilities in the Arctic. These sites usually have three diesel generators and a weeks worth of battery storage as back-up. Often the air inlets to the diesels are blocked by blowing snow. The load is automatically transferred to the battery banks. If the weather is still bad after a week and the battery bank is depleted, the facility will no longer operate. The communications company's technicians must then climb the mountain under blizzard conditions since the weather is too bad for helicopters. The diesel output and battery charge levels are monitored remotely by telemetry. With an operating wind turbine, service to the facility would not be necessary until a lengthy period of calm weather.

The most cost effective uses for wind turbines are at isolated facilities. In the long run, the land and high winds of once isolated areas will be utilized to install wind farms to provide bulk power supply to the major population areas.

Fig. 1. Energy Unlimited, Inc.'s North Wind HR2 Ultra Reliability DC Wind Turbine. It produces 2.2kw at 9 meters per second wind velocity and has a survival wind speed of 75 meters per second.

Fig. 2. Energy Unlimited, Inc.'s Windworker 10kw which is designed specifically for use with a synchronous inverter. It reaches rated power at a wind speed of 10 meters per second. The variable pitch rotor is 10 meters in diameter.

(American Refining/Energy Unlimited, 2 Aldwyn Center, Villanova, PA 19085, USA)

TOPIC E

Hydro, Solar, and Buildings

Appropriate Control Systems for Water Turbines

P. W. Agnew

Mechanical Engineering Department, University of Glasgow, UK

ABSTRACT

Water turbines are more difficult to govern than are some other prime movers. Hydraulic servo governors have been commonly used from the 1890's until recently, but they have certain limitations and are too expensive for use with very small machines. In future large hydro plants are likely to have electro-hydraulic governors. Small run-of-river plants that are not connected to the grid are most likely to have electronic load governors.

KEYWORDS

Hydro-electric; governor; automatic control; small; economic.

EARLY HISTORY

Water power came before wind power: but so far as is known the first centrifugal governor was one that was applied to a windmill by an 18th century English mill-wright called Lee. John Rennie, when he was an employee of Boulton and Watt, adapted Lee's design for use with a steam engine, and this is the device that became known as Watt's governor. Watt never claimed to have invented it.

It was not long before people attempted to apply governors to water wheels, but there were certain problems to be overcome. The most obvious problem, was that the output power of the governor had to be increased. In order to control one of Watt's low pressure steam engines, the governor operated a simple butterfly valve: the amount of energy required to open or close this valve was very small, so that it could be worked by a fly-ball mechanism of moderate size. In order to control a water wheel it was necessary to operate a sluice gate: a fly-ball mechanism could only do this with the help of what we would now call a servomechanism or mechanical amplifier, but was then known as a relay or cascade.

WATT'S GOVERNOR

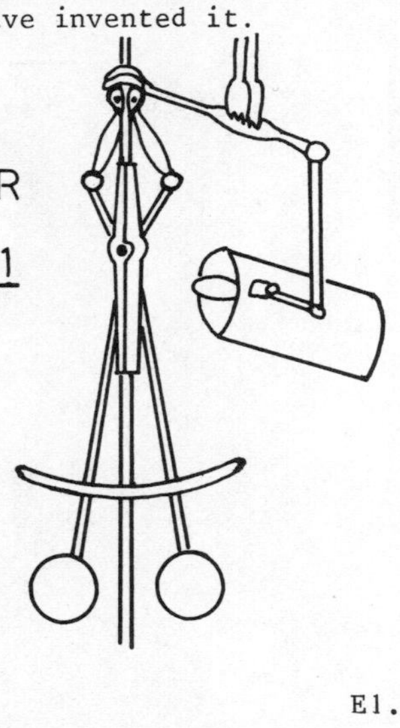

FIG 1

Much skill and ingenuity went into the design of these mechanical servos, and as water turbines came into use, from the 1820's onwards, very similar devices were installed in order to control them. (Like sluice gates, water turbine control mechanisms usually require quite a lot of operating power). Though they were certainly installed, it is hard to say how often these governors were actually used. Very often, instead of holding the speed steady, they caused it to oscillate or "hunt". Curing this trouble required not only improved servo design, but also mathematical analysis of the system as a whole in a manner that was understood by very few engineers in the 19th century and that did not become commonplace until the middle of this century.

As soon as suitable generators became available, which was round about 1880, hydro-electric plants began to appear. Quite a lot of small plants were built to supply country houses or rural estates: for example, the first hydro-electric plant used for lighting was at Cragside House, Northumberland (the owner of Cragside was a friend of Swan, one of the inventors of the electric light bulb). The plant at Pollok House, Glasgow, which was installed at the turn of the century, has a mechanical servo governor: the fact that it was apparently made inoperative at an early date, by tying it up with wire, suggests that it did not work very well. Both these houses are open to the public, though at neither is the hydro plant normally on show.

Plant Characteristics

The difficulty of governing hydro-electric plants is related to the manner in which the reservoir, pipeline, turbine, generator and electrical load are connected together and react on one another. The principal cause of the trouble is the relation between the pipeline and the turbine, and this is least difficult to describe and to analyse in the case of an impulse turbine, such as a Pelton wheel, that is controlled by means of a spear valve (see Fig. 2).

TWO-NOZZLE PELTON WHEEL

FIG 2

If we want an increase in power we increase the nozzle area in order to let more water through. But the water in the pipeline has inertia, so that we cannot instantaneously change the quantity of water flowing: the intial effect of a sudden increase in area is that the same quantity of water continues to flow but through a larger orifice, and therefore at lower velocity. There is therefore a transient drop in the power, and it takes a certain time for the flow to increase and for the power to stabilise at its new higher level. This is illustrated in Fig. 3. The peculiar nature of this transient response gives some idea of the problem involved in controlling water turbines: in order to describe the problem more precisely it would be necessary to use a more mathematical approach than would be appropriate in this paper. (Agnew, 1974).

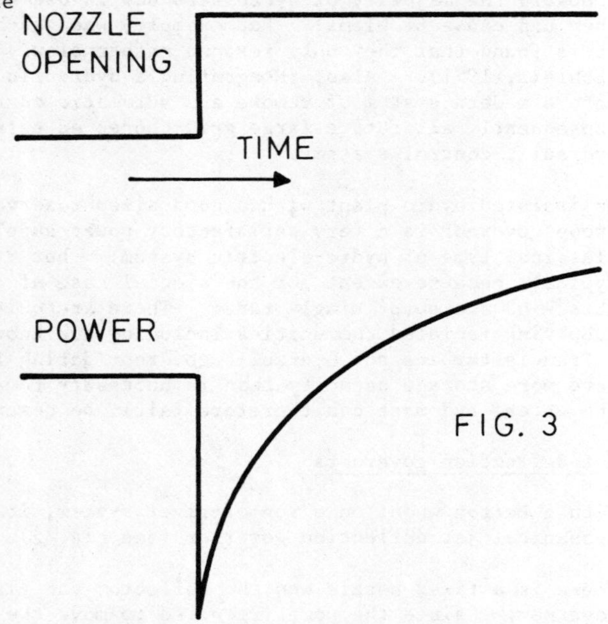

FIG. 3

"Temporary droop" governors

The first satisfactory solution to the problem, in the form of the hydraulic servo temporary droop governor, appeared in the 1890's and gradually came into general use. The expression "temporary droop" needs some explanation.

The "droop" of a governed prime mover is the drop in speed between no load and full load: it is often expressed as a percentage - 5% droop means that the speed changes by 5% between no load and full load (see Fig. 4). A "temporary droop" governor behaves as if it has a high value of droop (say 30%) for rapid transient changes in speed: for slow changes it displays a relatively low value (say 3%) of "permanent droop".

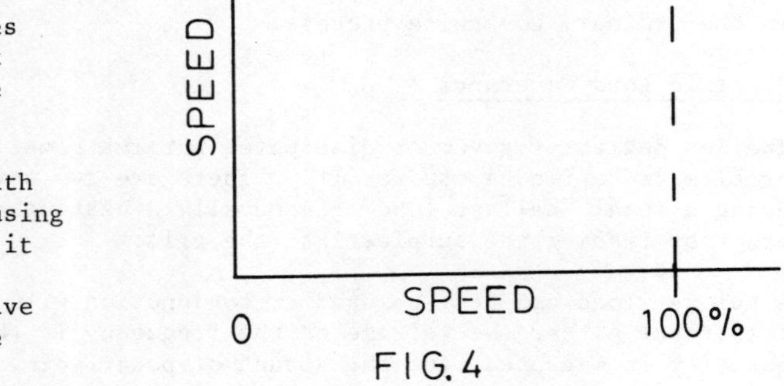

FIG. 4

In order to govern a turbine with the behaviour shown in Fig. 3 using a simple proportional governor it may be necessary to use a very large droop, which would not give adequate speed control, because a governor of this type with a lower value of droop would cause instability (oscillation or 'hunting").

By using a temporary droop governor it is possible to obtain both a tolerable quality of stability and a practically acceptable value of the permanent droop.

Probably the majority of hydro sets now in use have hydraulic droop governors, but they can cause problems. For example, when such sets are connected to the grid, it is found that they only respond rather slowly to changes in grid frequency (Schleif, 1971). Also, integrating a hydraulic governor of the traditional type into a modern system of remote and automatic control can be a complicated business. Consequently all future large grid-connected sets are likely to have electro-hydraulic control systems.

An isolated hydro plant with a good sized reservoir and a hydraulic temporary droop governor is a very satisfactory power supply: it could be described as the classical type of hydro-electric system: but it would be wrong to regard it as typical, because except for the special case of aluminium smelters, systems of this kind are surprisingly rare. There are quite a lot of small hydro plants supplying isolated communities including one known to this author where there are a Francis turbine and hydraulic governor dating from the 1890's, but they seldom have more storage capacity than is necessary to allow the gravel to settle out of the water, and most can therefore fairly be describes as run-of-river plants.

Jet-deflection governors

With a Pelton wheel on a run-of-river system, it is possible to use a simple mechanical jet-deflection governor (see Fig. 2).

There is a fixed nozzle and the deflector vane is operated by a simple fly-ball governor: since the power required to move the vane is very small, this arrangement works quite well. Since the rate of flow in the pipeline is not affected, there is no stability problem. If the flow available is not enough for full power all the year round, either a hand operated spear valve can be used or a large ("winter") nozzle can simply by unscrewed and replaced by a small ("summer") one. The arrangement shown in Fig. 2, with spear valves as well as deflector vanes, would only be used on fairly large installations.

Hydraulic servo governors are expensive, and the use of a simple jet deflection governor can greatly reduce the cost of a small plant. Installations of this kind were particularly popular in the rural parts of Wales, up to about 1950 (Ashby, 1979). The same type of governor can be used with a Turgo wheel, but not with a Francis or propellor turbine.

Most Pelton wheels with conventional spear valves and servo governors have jet deflectors as well, but these are solely for emergency use and for obtaining rapid shut-downs without undue pressure rise in the pipeline, and play little or no part in the ordinary governing process.

Electric Load Governors

The jet deflector governor dissipates surplus power: a better solution to the problem is to find a use for it. There are two ways of doing this: one is by using a local "ballast load" - generally a heating system of some kind - and the other by feeding the surplus into the grid.

A ballast load can best be used in conjunction with an electronic load governor. This takes either the voltage or the frequency as the controlled variable: this quantity is measured, and the amount of power going to the ballast load is adjusted so as to keep it substantially constant. It is the advent of Silicon Controlled Rectifiers (Thyristors) for the instantaneous control of a power measured in kilowatts that has made this type of governor possible. In its simplest form it can only be used with a.c. systems: but this is no disadvantage since d.c. is hardly ever used nowadays for general purpose power supplies. If the ballast load takes the form of an immersion heater in a tank connected to a central heating system, each kw delivered continuously over 24 hours can be made to give 2kw of heat for 12 hours per day.

A load governor and ballast load, unlike a jet deflector governor, can be used with any type of turbine - for example a multi-jet Pelton wheel, Francis, or Propellor turbine. There is no doubt that propellor turbines of one kind or another are best for low head sites, and that many of the best small sites are run-of-river sites with a low head. The availability of suitable electronic load governors has made it practicable to use these sites. Seasonal variations in flow can be allowed for either by having two turbines, one large and one small, or by having a single machine with a variable pitch propellor. The use of a propellor whose pitch can only be adjusted when stationary makes the system appreciably cheaper and simpler, and is no great disadvantage for the operator.

Small Grid-Connected Plants

The other way of using surplus power is to feed it into the grid. With a small plant, this can most conveniently be done by using an induction generator. No governor is then needed. It is not necessary to synchronise such a plant - you simply switch it on.

With a run-of-river system, an automatic reservoir level control system can be used to ensure that the maximum possible amount of power is generated and no water is wasted. Since it requires a remote reservoir level sensor and a servo-operated turbine controller, such a system is bound to be fairly expensive, and is likely to be used mainly on the larger grid-connected induction generators.

In conclusion, it seems to this author that it was a short-sighted policy which caused a lot of small hydro plants to be closed down as the grid spread out into the rural areas. If it had been a matter of policy to connect these plants to the grid and to encourage the building of new ones, some of our wetter and hillier rural areas could be net contributors of power to the grid instead of being simply a load on it. This could provide an additional source of income in some of the poorest areas of the country, and could also give the people in these areas a means of establishing their self-reliance and contributing something useful to the common good. Communities that come to depend on outside help and subsidies for all the essentials of life, are doomed. In many of our rural areas, small scale hydro power is simply waiting to be exploited, and modern technology has made it possible to exploit this power more economically than ever before.

REFERENCES

Agnew, P.W. (1974) The governing of Francis Turbines. Water Power, 26, no. 4

Schleif, F.R. (1971) Governor characteristics for large hydraulic turbines. U.S. Dept. of the Interior, Bureau of Reclamation, report no. REC-ERC-71-44.

Ashby, M. (1979) Water power in the Dyfi valley. Centre for Alternative Technology, Machynlleth, Powys.

Solar Cells in Scotland

F. Riddoch and J. I. B. Wilson

Department of Physics, Heriot-Watt University, Riccarton, Edinburgh, UK

ABSTRACT

Solar cells have been used since 1978 to power the telephone system on Soay off the South-West coast of Skye, and since 1975 to power radio repeater stations elsewhere in Scotland. This paper reviews the history and performance of these installations and looks at the prospects for solar cells in the overall energy supply pattern at these latitudes. While cells are already economic for specialised telecommunications applications, cost will continue to limit their domestic use.

KEYWORDS

Cost; batteries; supply pattern; northerly latitudes; telecommunications.

INTRODUCTION

Solar cells (photovoltaic diodes) convert sunlight directly to electricity. Cells are manufactured from a single thin slice of crystalline silicon and are commercially available mounted and wired into "arrays", whose efficiency per unit area is around 5%. Solar cell technology developed rapidly in the 1960's to meet the need for a light-weight, reliable power source for satellites. More recently, the energy problems of the 1970's have given new impetus to adapting this technology for ground-level use. Research is aimed both at bringing down the cost and at increasing the efficiency of solar cells and arrays. The high cost of arrays means that at present they can only compete financially with other energy sources in remote areas where grid electricity is not available, and especially where access to the site for maintenance of more traditional generating equipment is difficult and expensive. These are exactly the conditions which exist in some of the remote parts of Scotland today, where arrays are being used successfully to power isolated telecommunications equipment.

This paper reviews the applications of photovoltaics to telecommunications in Scotland and further, by considering the insolation data available, looks at the prospects for solar cells in the overall energy supply pattern at these latitudes. Solar cell arrays have been used since 1978 to power the telephone system on Soay off the South-West coast of Skye, and since 1975 to power radio repeater stations in various parts of Scotland.

SOAY TELEPHONE SYSTEM

Soay has a population of five adults and two children. The GPO equipment on the island consists of three telephones, a kiosk, a ten line electronic line changer and a VHF radio transmitter, which together comprise an electric load of 260 mA, 12 volts continuous, with peaks: an estimated 9.6 Amp hours per day overall. The equipment is powered by lead-acid batteries which require regular recharging. The recharging was done locally using a petrol generator on the island, but in 1975 the Post Office decided to review the options for meeting this load. The main problem lay in finding generating equipment, of suitably rated output, which required the minimum attention as it was decided that if possible the Post Office should cease to rely on local help to power the system. A prime objective had to be to minimise the man-hours of trained personnel required for Soay as the real costs associated with travelling to the island and servicing equipment were escalating. Several possibilities were considered:

(a) Petrol generator. This type of equipment would certainly require some degree of manual operation to function well; maintaining fuel and water level, and checking operating times and securing fuel. The cost was estimated at £300.

(b) Windmill. The Telecommunications Board had in the past experimented with a wind generator, but found that at its test site near Haddington, East Lothian, the model chosen failed to withstand gusting. However, as a windmill had been used for power at Stornoway in the past, this option was investigated and a Windco Windcharger was tested at Inverness. Although the overall output from the windcharger was adequate for the load (the wind speed on Soay is greater than 9 mph for 66% of the time), the output could not be smoothed sufficiently to meet the needs of the equipment. Lingering doubts about tethering and the mills resiliance to gusting remained, and regular maintenance seemed advisable. The Windco Windcharger cost: £200.

(c) Fergusson cells. These are a type of battery used by the Post Office in remote stations. The Fergusson cell is a robust primary cell (i.e. it is not rechargeable) and two sets would have been adequate for Soay. Annual replacement would have been necessary. Fergusson cells would cost £720/annum.

(d) Solar cells. Using Stornoway (Lewis) insolation data, an estimated 2 panels rated at 13.5 V and 3.0 Amps in AM1 (ideal) conditions with an area of 1.0 m^2 would be needed to meet the load. The panels would be used to recharge one 500 Amp hr battery and this would give sufficient storage to maintain the supply through 50 days without cell output. Once installed the panels require a minimum of maintenance and there are no fuel costs, but the initial capital cost would be high, at around £1,000.

(e) No active equipment. A mains cable could be laid over the three miles separating Soay from mainland Skye at Elgol. This would cost an estimated £30,000.

Cost ruled out the cable option. Of the rest, those with lowest maintenance requirement were most favoured and solar panels in this context were very attractive. Solar panels were already being successfully used by the GPO on a pilot scheme, to power an emergency telephone in Norfolk.

Three panels were installed on Soay in 1978, mounted on a free standing platform beside the two huts which house the radio equipment and batteries. The cell output is connected via a voltage regulator to the batteries and an ampere hour meter has recently been installed to help monitor the panel performance. After 18 months the GPO have had no problems with the panels and are now considering installing two similar links: one at Badanloch and the other connecting Tomdouhn and Kinloch Hourn. This latter installation will cost the GPO roughly £650 for one solar panel, twelve batteries and a voltage regulator at 1980 prices.

CIVIL AVIATION AUTHORITY (CAA), HILL-TOP REPEATER STATIONS

In 1975 the CAA decided to use solar arrays to recharge the battery banks of some of their radio repeater stations (Tuke, 1975). The Authority are responsible for coastguard communication and to give good coverage over the west coast of Scotland had installed repeater stations on Mull, Hoy, Islay and in the north of Skye. The stations are powered by batteries and constitute a load of 5 amp hours/day. Housed in specially designed weather-tight cases, the transmitter delivers 10 W output for 24 V 1 amp input. Replacing and recharging the batteries on such remote hill-top sites posed a considerable problem. The Authority considered three options for reducing maintenance costs on these sites: a windmill, atomic batteries, or solar cells. The windmill was considered unreliable in the prevailing wind and weather conditions, while the cost and environmental constraints on the atomic cells were prohibitive. Solar cell arrays were installed in 1975-76 and these arrays provide valuable information as to how well panel technology can withstand our climate. Although the CAA found the cells performed adequately, the arrays developed unexpected problems after the first three years of operation. One seasonal problem encountered at all the sites was due to ice forming on the panel surface. A solar panel generates heat during operation, enough to hold it a few degrees above ambient. Immediately darkness falls the cells cool, become very good radiators and ice can form on the panel surface. The opacity of the ice can stop the cells generating when the sun rises and the cells will remain off until the ice melts. With time, the panel encapsulant on the repeater stations turned yellow, which gradually decreased the cell output, while water got into and damaged some of the panels. The yellowing effect was a common problem with early panels and is now known to be caused by UV radiation. The lack of resistance the panels showed to water is also a recognised problem with certain forms of encapsulation. Early panels were sealed simply by embedding the cells in silicone rubber with or without a top glass covering. As the hill-top repeater sites showed, this does not give a weatherproof finish. The most recent suggestions (Salles) on encapsulation are much more rigorous. Heat and pressure treatment should be used to seal the cells between glass plates avoiding silicone rubber altogether because of its cost and other disadvantages. By the end of 1980 the CAA's four sites will have been dismantled. The Authority has joined with a wider group of radio users to extend its coverage and this combined with an increased amount of radio traffic has made the repeater stations no longer necessary.

SOLAR CELLS IN SCOTLAND

The factors limiting the use of solar arrays at any latitude are:

(1) the mean solar irradiance, which sets the theoretical upper limit on energy available from this source;

(2) the efficiency and cost of solar cells and batteries.

TABLE 1 Mean global daily-radiation on a horizontal surface. The Lerwick data is averaged over 26 years, Aberdeen over 10 years, Dunstafnage over 8 years.

	LERWICK (watt h $m^{-2}day^{-1}$)	ABERDEEN (watt h $m^{-2}day^{-1}$)	DUNSTAFNAGE (watt h $m^{-2}day^{-1}$)
January	247	366	458
February	791	966	1011
March	1660	1963	2097
April	3210	3269	3910
May	4054	4074	4456
June	4683	4839	5107
July	4088	4307	4640
August	3255	3621	3648
September	2115	2264	2419
October	1006	1251	1319
November	357	593	605
December	158	295	359

Insolation. Table 1 shows the mean daily radiation for each month of the year recorded at the Meteorological Office stations of Lerwick, Aberdeen and Dunstafnage (near Oban). These figures are for global radiation on a horizontal plane and represent the daily average over a number of years. When mounting a solar array the angle of tilt should be close to the angle of latitude, although in high latitudes to take advantage of what little insolation there is in winter, the best results may be achieved at slightly greater angles. Using the Met. Office data, the output of one square metre of solar panels (i.e. two panels each rated at 13.8 V and 2.0 amps in AM1) lies between 120 watt hr/day and 300 watt hr/day from the month of April until September. This energy is sufficient to completely recharge a standard (12 V/50 A hr) car battery in two days in mid-summer or 5-10 days in spring/autumn from 1 m^2 of solar panel. The calculations are accurate for a horizontal plate and hence pessimistic for a properly angled array. For the winter months the panel output is at its lowest and it could take over twenty days to recharge the battery. Because of this large annual fluctuation it is always necessary to incorporate batteries in the energy supply system to maintain a fixed output in the winter months and not simply to provide a night-time supply. Additional batteries are recommended over and above this to allow for exceptionally poor weather conditions.

COST AND EFFICIENCY OF SOLAR POWER SYSTEMS

Solar cells of the type used in the applications described are p-n junction, crystalline silicon cells. Their manufacture is costly in terms of energy, materials and required sophisticated engineering. In Britain today solar energy in a useable form (i.e. a panel) costs ∿ £350 for 31 peak watts.

Considerable research is in progress, especially in the USA, directed towards reducing the cost of solar cells both by improving traditional techniques and investigating new materials. Individual cells are 15-17% efficient but overall panel efficiency is less due to wasted space between cells. New types of silicon are being investigated including amorphous silicon (α-Si), which is the subject of research at the Heriot-Watt and Dundee Universities in Scotland. One advantage of α-Si is that it should be relatively cheap and easy to make and could be used to cover large areas continuously.

Batteries are another area of important research for solar power systems. If a guaranteed supply is required, or even a night time supply battery storage is an essential part of a solar energy system.

TABLE 2 Characteristics of selected batteries (likely achievable performance). "Advanced Battery Development" J. Jensen, International Power Generation 2, 1979.

	Lead Acid	Nickel Cadmium	Nickel Iron	Nickel Zinc	Iron Air	Sodium Sulphur
Electrolyte	H_2SO_4	KOH	KOH	KOH	KOH	KOH
Voc	2.05	1.35	1.37	1.71	1.71	1.8-2.1
Energy efficiency (charge-discharge) %	75	70	<60	75	40	70-75
Energy density (Wh/m^3)	70	60	100	140	80	170
Life (to 80% discharge) in cycles)	500	2000	2000	350	200	2000
Cost: Capital £/kWh	25	250	65	30	23	30
Cost: Running £kWh cycle	0.05	0.2	0.03	0.1	0.11	0.02

Table 2 compares the properties of rival batteries. Apart from being cheap a good system battery would also have:

(a) proven ability to withstand long charge/discharge cycles;
(b) low self discharge rate;
(c) a low power/weight ratio;
(d) require little maintenance.

The lead acid battery still comes closest of all to meeting these requirements.

In the applications discussed here requiring a low wattage autonomous power supply, solar panels charging lead-acid batteries proved to be the most economic means of supply. However this is a rather specialised class of energy supply problem more often encountered in communications and transport than in domestic situations. The only comparable supply an individual might need is the farmer requiring an autonomous supply for an electric fence and solar array manufacturers are just beginning to investigate this possible market.

Domestic power requirements are usually greater than those discussed and the alternative sources of supply in the kilowatt range of output more competitive. Table 3 indicates the area of module, battery storage and approximate capital cost of meeting various domestic loads from solar arrays at present. For

TABLE 3 Solar array battery requirements to power specified loads, in Lerwick area.

	Load Amp hr /day	No. Modules (13.5V,2.0A)	30 day Battery Storage (Amp hr storage)	Approx. Capital Cost
Radio Equipment 10W used 2 hrs/day	5.0	2	∿ 150	£750-£930
T.V. (Black & White) 16 watts for 5 hours daily: 12.V.D.C.	6.6	2	∿ 198	£750-£930
Fluorescent lighting 4 x 13 watt lights 2 x 8 watt lights (5 hr /day) 12 V.D.C.	28.33	7	∿ 378	£2,600-£3,255
6.0 KVA diesel generator 0.4 litre /kWh				£5,000

comparison the capital cost of a 6 KVA diesel generator and its rate of fuel consumption are indicated. The area of panel required for these applications is not prohibitive although the estimated cost of the system is too high for most individuals. However, if the cost per peak watt of panels and cells can be reduced in the coming years arrays could be used to meet specifically electric loads in a mixed fuel domestic supply.

ACKNOWLEDGEMENTS

In researching this paper I was greatly assisted by both the Scottish Telecommunications Board and the Civil Aviation Authority. I would like to thank particularly Mr T Kirkwood and Mr N Penny of S.T.B. and Mr H Graham of C.A.A. for their cooperation.

The views expressed here are our own and do not necessarily reflect the policy of either of the above mentioned authorities.

REFERENCES

Salles, Y., J. Agnet and A. Desombre. Study of Sandwich Cell Encapsulatra, R.T.C. Caer, France.

Tuke, J.B. (1975). Wireless World, May.

Solar Heat for the Highlands and Islands of Scotland

A. W. K. MacGregor

Napier College, Colinton Road, Edinburgh, Scotland

ABSTRACT

This paper examines some possibilities for using solar energy to provide heat in the Highlands and Islands of Scotland. It is concluded that there is considerable potential for the use of solar heat because of the existence of (a) a large demand for hot water in summer, (b) a high proportion of the housing stock suitable for passive solar retrofit, and (c) a relatively favourable climate for solar space heating.

KEYWORDS

Solar heat; Scotland; Highlands and Islands; climate; building design.

INTRODUCTION

Less than a century ago the economy of the Highlands and Islands of Scotland was based almost entirely on solar (or solar induced) indigenous and renewable forms of energy. Now, apart from hydro-electricity, it is largely dependent on imported fossil fuels which are rapidly becoming scarcer and more expensive. For future survival, alternative, non-depleting sources of energy must be developed before the fossil fuels are entirely consumed. This paper examines some possibilities for using solar energy to provide heat for buildings.

ENERGY USE

A high proportion of total energy consumption is used for the production of low temperature space and water heating. For example, in the domestic sector it is estimated that about 85% of final energy consumption is used for these purposes[1]. The remaining 15% is used for cooking, lights and appliances and in most situations this can be supplied most conveniently and at moderate cost by electricity. For most energy users the major part of their fuel bill arises from the need to keep their buildings warm and to provide a supply of hot water.

SOLAR ENERGY

There are many ways of collecting and using solar energy. Indirectly, it provides the motive power for wind turbines, wave machines and hydro plant. Directly it can be converted into heat, electricity, or a chemical/biological fuel. Conversion to heat is particularly attractive because (a) the efficiency of conversion is relatively high, (b) cost per unit area of collection surface is relatively low and (c) a high proportion of the energy we use is required in the form of heat. The temperature at which the heat is required determines largely the type of solar collector which will be used. For the higher temperatures which are required for operating power stations, refrigeration plants or solar furnaces, more expensive and complex focussing collectors are required. These can use only the direct rays from the sun, rather than the diffuse radiation from the whole sky, and are therefore not well suited to the Scottish climate with its high proportion (> 50%) of diffuse solar energy. Fortunately most of the heat we use is at lower temperatures (< 50°C) and this can be provided by cheaper and simpler stationary flat-plate collectors.

SOLAR WATER HEATING

The outline of a typical flat-plate solar water heating collector is shown in Fig. 1.

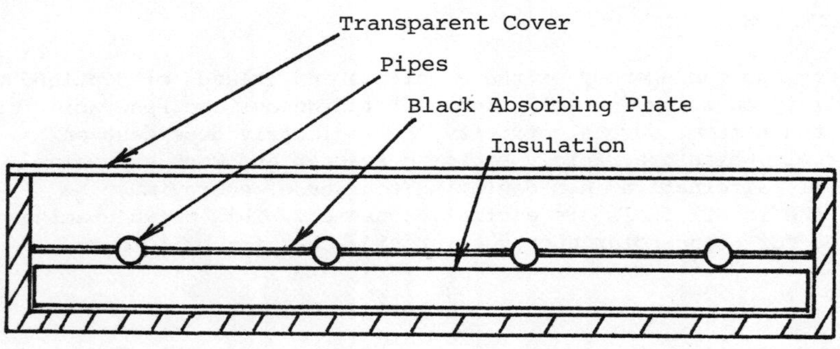

Fig. 1: Cross-section of Flat Plate Solar Water Heater

A black painted plate, with water passages incorporated or attached, lies under a transparent cover and is backed by a layer of insulating material. Solar radiation passes through the cover, is absorbed by the plate, conducted to the water and removed by circulation of the water to a storage vessel. There are many variations on this basic design, including choice of materials, number of covers, method of water circulation and type of absorbing surface. Most commonly, the plate is made of black-painted copper or aluminium. The water passages are copper, a single sheet of plain glass acts as the cover and circulation is by pump to a heat exchanger coil in a pre-heating storage tank. In the Scottish climate a solar water heating system is normally used as a pre-heating fuel saver rather than as a complete replacement for conventional plant. More detailed information on the design of systems can be obtained from references 2, 3 and 4.

The main use of solar water heaters is for the supply of domestic hot water. In the Highlands and Islands a major application could be in hotels, hostels, chalets and camping sites where there is a heavy demand for hot water, specially in the summer months. Other applications specially well suited to solar heating are fish farms and swimming pools, where a large quantity of water has to be heated through a small temperature rise. The fuel savings which can be achieved by solar water heaters depend largely on the particular application, particularly the pattern of use and the type of fuel being replaced. However, an annual useful energy yield of between 300 and 500 kW hours for each square metre of collecting surface could be reasonably expected. Installed costs vary even more widely, ranging from about £30/m^2 for a large "do-it-yourself" system to about £300/m^2 for a small professionally installed system.

An interesting new development is the "clip-fin", an aluminium fin which is shaped so that it clips closely over an ordinary water pipe and transforms the pipe into part of a flat plate solar collector. This allows the constructor to assemble a solar collector of any size on site, using conventional plumbing methods and avoiding the purchase of the expensive, specialist collector boxes which are presently on the market. The "clip-fin" method was used to build the 18 m^2 solar water heating system on the Island of Iona, described in another paper.

SOLAR SPACE HEATING

In most buildings the energy used for space heating represents the single biggest item on the energy bill. Since the heat is required at a moderate temperature level, a feature well suited to solar systems, and the amount of heat required over the year is usually a fraction of the solar energy incident on the building's outer surfaces, then this would appear to be a very suitable application for solar energy. On the other hand, the solar supply is intermittent and is six months out of phase with the demand for heat. Most of the early work on solar space heating systems was done by engineers and they naturally adopted the approach of using engineering hardware to collect, store and distribute the solar heat. In effect, this meant replacing the boiler or furnace of a central heating system with a large array of water (or air) heating collector panels on the roof and adding a large water (or rock) tank to store heat for periods ranging from hours to months. It is technically possible to design a system which will provide 100% of a building's heating requirements from solar energy but the capital cost of providing a collector array, covering the complete roof area, and a storage volume approaching the volume of the building itself makes this approach impractical at present. With existing technology a reasonable target is about 50% solar heating. Recent developments such as lower cost "clip-fin" collectors and more compact change-of-phase chemical heat stores - together with the continuing real increase in fuel prices - could make it worthwhile to aim for a higher solar contribution in future designs.

In the Highlands and Islands, the main application for "active" solar space heating systems, i.e. those involving separate collectors, storage tanks and pumped circulation, is likely to be in new buildings of timber-frame construction where the light-weight building fabric cannot be used as a storage element, but large scale collectors can be integrated into roofs as part of the original design. An alternative approach for using solar energy to heat buildings is the "passive" method where the fabric of the building itself acts as a combined collector, store and heat emitter. In many respects this is a re-discovery of the merits of traditional, climate oriented architecture, but the availability of modern materials has widened the scope considerably. Recent work, particularly in the

U.S., indicates that it is possible to achieve results comparable with fully "active" solar systems but at much lower cost and with less complexity. The basic principle of "passive" solar design is that the external surfaces of the building are chosen so that they absorb or transmit solar energy and the structure of the building is made sufficiently massive to store the collected energy over short periods. There are many variations on this theme.

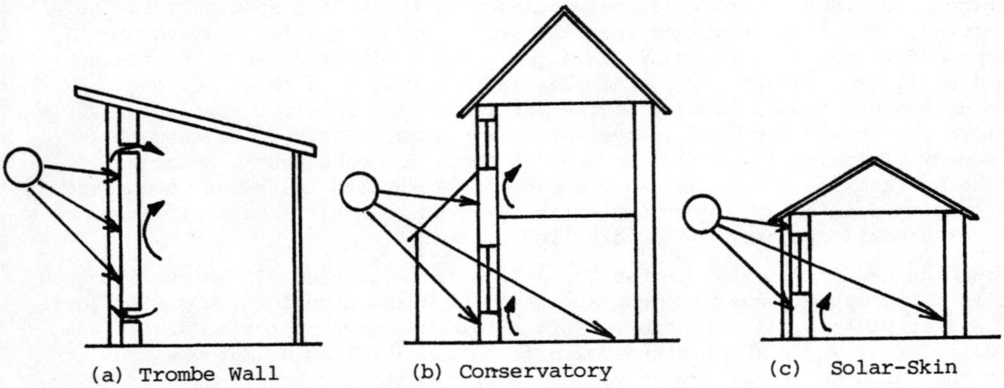

Fig. 2: Passive Solar Heating Methods.

One of the most effective is the Trombe Wall (Fig. 2a), developed in the Pyrenees, in which a layer of glass covers a dark painted, south facing, concrete wall with air vents to the room space. The solar heat absorbed at the outer surface of the concrete wall is transmitted to the room both by convection through the slots and by conduction through the concrete. The optimum thickness of the wall is about 350 mm. An additional refinement provides cooling as well as heating. A major disadvantage of the full Trombe design is the loss of light to the building. This is overcome in the conservatory or lean-to glasshouse (Fig. 2b) attached to the south wall of a house. It provides most of the heating benefits of the Trombe wall as well as additional space for living in and growing plants. This could easily be applied to the traditional stone built house of the Highlands and Islands. If space and money are scarce, a third alternative is the "solar skin" (Fig. 2c) which is merely a translucent cladding of glass or plastic which is fixed directly to the external surface of the wall over a small air gap. Whether intended or not, every house is a "passive" solar collector to some extent, with solar energy being transmitted through the windows or absorbed by the walls. In the Scottish climate these incidental solar gains should be maximised by locating most of the windows on the southern aspect of the house and using a dark colour if the house is to be painted. The traditional solar-rejecting white paint of many houses in the Highlands and Islands is better suited to North Africa than to Scotland.

CLIMATE

It is commonly believed that Scotland's climate, particularly in the Highlands and Islands, is not well suited to solar space heating. On the contrary, however, it has been shown[5] that the increase in the length of the heating season as one moves North in the British Isles, allows a much higher proportion of the solar energy available throughout the year to be usefully employed in saving fuel. In other words, an identical solar system attached to the same house in Aberdeen and in London will save more fuel in Aberdeen and is therefore more cost effective at that location. Fig. 3 illustrates the variation in solar supply and heating demand at these two places, and Fig. 4 shows the consistently improving trend of cost effectiveness with increasing latitude.

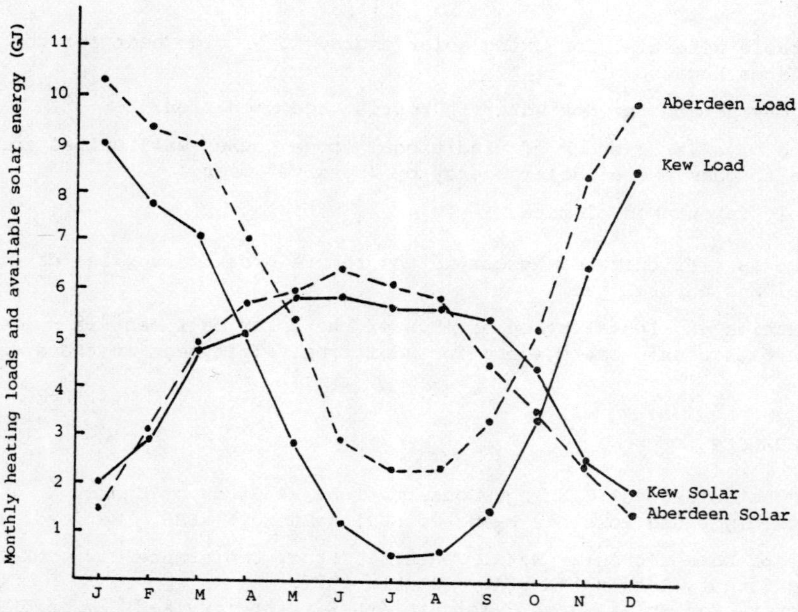

Fig. 3: Heating loads and available solar energy at Aberdeen and London, for a house with specific heat loss of 300 W/K and a 68 m² vertical collector of 30% efficiency. The useful solar energy in any month is limited by the available solar energy or the heating load, whichever is the lesser.

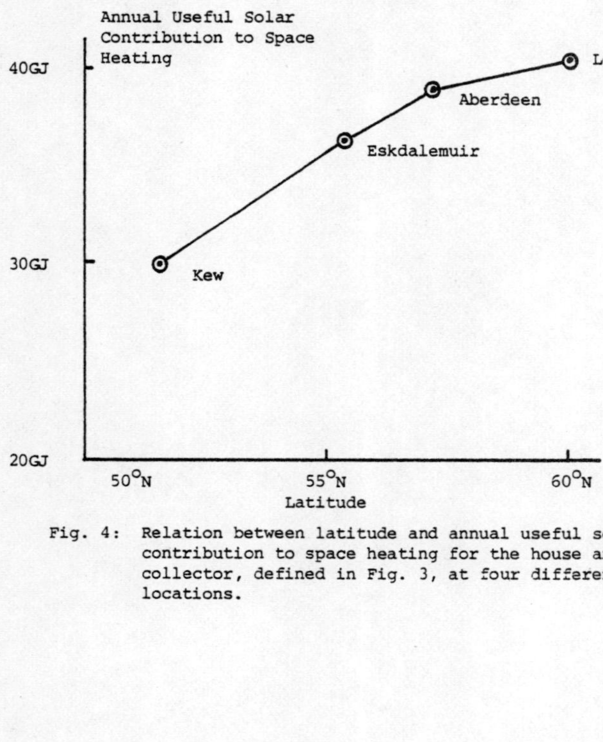

Fig. 4: Relation between latitude and annual useful solar contribution to space heating for the house and collector, defined in Fig. 3, at four different locations.

CONCLUSIONS

There is considerable potential for using solar energy to provide heat for the Highlands and Islands because of:

(a) the high summer demand for hot water in tourist accommodation.

(b) the existence of a large stock of traditional stone houses well suited for modification to make use of solar energy by "passive" means.

(c) the relatively favourable climate.

An added incentive is difficulty of access in the remote area to supplies of fossil fuels such as coal and natural gas.

Since the construction and installation of much of the solar equipment is suited to local skills, there is considerable scope for additional employment in the area.

REFERENCES

1. Building Research Establishment. Energy Conservation: A Study of Energy Consumption in Buildings and Possible Means of Saving Energy. BRE. Watford. 1975.
2. Practical Aspects of Domestic Solar Water Heaters. Proc. Conference C12. UK-ISES, 19 Albemarle Street, London W1X 3HA. October 1977.
3. Flat Plate Collectors and Solar Water Heating. 4th Edition (revised) 27 pp.UK-ISES, 19 Albemarle Street, London W1X 3HA. April 1977.
4. K McCartney. Practical Solar Heating. Prism Press 1978.
5. A W K MacGregor. The Potential for Solar Space Heating in Scotland. Int. Journal of Ambient Energy. Vol 1, No 3, July 1980.

Solar Water Heater on Iona

J. Carrie

Church of Scotland SRT Project, 121 George Street, Edinburgh, UK

ABSTRACT

A low cost Solar Water Heater has been installed at a Youth Centre on the island of Iona. Although frost damage has occurred and other difficulties experienced the collector is performing satisfactorily. The conclusion is that similar 'Do-it-yourself' Solar Water Heaters are an economic proposition for Rural and Island Communities.

INTRODUCTION

Close to Iona Abbey is the Iona Community's Residential Youth Centre which each week accommodates 50 young people during a season which lasts from April to October. This Centre has two domestic hot water systems, one serving the kitchens and staff accommodation, the other the dormitories. The two systems were electrically heated.

A survey in May 1977 showed that about 650 litres of water at 60°C were being used each day in the kitchens and staff accommodation. It therefore appeared likely that some 7,700 kWh of electricity was being consumed each season to warm this water alone.

PROPOSED SOLAR ENERGY PROJECT

The Iona Community with the Church of Scotland's Society, Religion and Technology Project set up a working party charged with installing a solar water heater for the kitchen and staff block which would save at least half of the electricity previously used to heat the hot water and pay for itself within five years.

In order to save half the electricity, a collecting area of about 16.5m^2 was required, on the assumption that the solar water heater would utilize about a third of the 700 kWh of solar energy that each square metre of Iona generally receives within the April-October season. In order to pay for itself within five years, the cost of the system had to be kept under £450 - at 1978 prices.

THE LOW COST SYSTEM.

That such an inexpensive water heater was built in June 1978 is due in part to the work being carried out by volunteers, and in part to the expert help of Kerr MacGregor of Edinburgh's Napier College. MacGregor (1978) was developing a system of bonding grooved aluminium plates to copper pipes with an araldite adhesive. This technique was applied on the grand scale on Iona.

Figure 1 shows the solar water heater on the roof of the Youth Centre. The orientation is 20° west of south and the tilt is 35°. This angle of tilt was selected since there is no requirement to collect solar energy in the winter. The design of the heater was determined by the roof space available and the size of the glass selected, which is the inexpensive 3mm 'Dutch Light' horticultural glass. There are eighteen glazed sections and a smaller unglazed section which houses the manifolds. The effective collecting area is $17m^2$.

The frame is 70 x 45mm redwood and is fastened to the roof by coach bolts. The astragals are mounted at 755mm centres, and each has eight holes cut in it at 170mm centres, through which pass lengths of 15mm copper pipe. The astragals were mullion cut for ease of assembly and maintenance. The back of the collector is weatherboarding over builders' paper, and there is felt flashing.

Fig. 1 The Solar Water Heater on Iona.

At one end of the collector are two manifolds made from 28mm copper pipe. The manifolds are connected to each other by four long loops of 15mm copper pipe, the loops running the entire length of the structure. The various joints are made from standard capillary fittings, and there are drains and air cocks.

The collector plate was made by gluing grooved 22 gauge aluminium plates to the 15mm copper pipes with Ciba-Geigy Araldite AV 138 with Hardener HV 998. Some 144 plates were used, each measuring 710 x 170mm, with a 15mm groove along the centre. The fins were prepared at Napier College. Others using this technique could either make their own plates or else order them from a suitable firm.

The collector plate was painted with ordinary matt black paint and insulated with aluminium foil, old blankets and large quantities of reject Phenexpan insulation board. The construction of the water heater is illustrated in Fig. 2.

Fig. 2 Construction of solar water heater.

A. 15mm copper pipes.
B. Insulation board
C. An old blanket for additional insulation
D. Aluminium foil
E. Aluminium fins.

ADAPTING THE EXISTING HOT WATER SYSTEM

The existing system had a 750 litre roof tank feeding cold water into the hot water cylinder. Since this tank only served the hot water system, it was decided to use it as a pre-heat tank. Water had been supplied to this tank by a ballcock, but since a ballcock should not be used in hot water and to allow the cold water to enter at the bottom of the pre-heat tank, a small plastic tank with a ballcock was installed beside the existing tank, as shown in Fig. 3.

Iona water is 'soft' so a 'direct' system with the water in the tank circulating through the solar collector was possible, and selected. Water is drawn from the bottom of the tank to the solar collector through a 28mm copper pipe with a pump and non-return valve. The warmed water is returned to the tank by 28mm copper pipe and a system of baffles made from old plaster piping. The baffles were intended to reduce turbulence and allow the warmest water to gather at the top of the tank. The draw-off to the hot water cylinder was extended to the top of the pre-heat tank, where the water should be warmest. To ensure that the whole capacity of the tank remains available, a small leak was provided between the extension and the bottom of the tank.

An electronic controller activates the pump whenever heat can be collected. Hewlett-Packard Ltd. designed and donated the controller used on Iona. Other suitable controllers are available commercially.

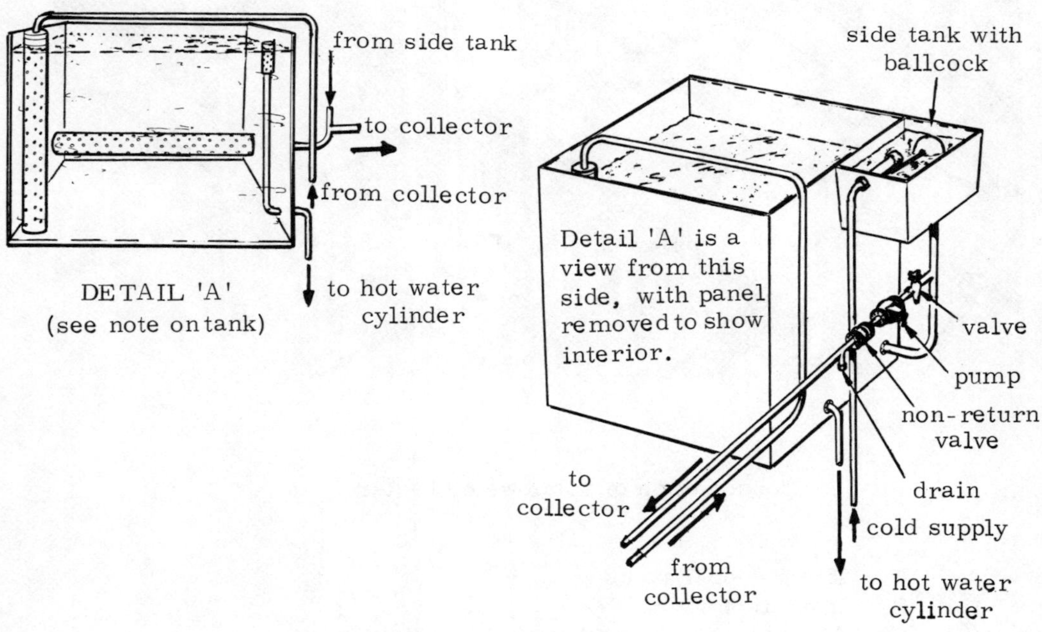

Fig. 3 The pre-heat tank.

THE COST OF THE SYSTEM

An analysis of the costs actually incurred on the Iona Solar Water Heater will not be helpful to others since most of the material was donated. However, an attempt was made to indicate how much a similar system might cost. For this costing (Carrie 1978) it was assumed that a suitable sloping roof was available (to eliminate the need for a complex frame and expensive back), but that a pre-heat tank had to be purchased (which was not the case on Iona). Table 1 shows that in 1978 a similar $17m^2$ might have cost about £475, or £199 plus £16.24 per square metre of collector surface.

TABLE 1. The cost of an equivalent collector in 1978.

Cost of Basic Collector

Glass (18 panes of 3mm Dutch Light horticultural glass)	£ 31
Aluminium (8 sheets of 22g Aluminium, 8ft. x 4ft.)	£ 55
Adhesive (2 kg AV 138 with 0.8 kg HV 998)	£ 23
Copper pipes (120m of 15mm pipe and fittings)	£ 51
Insulation board and aluminium foil	£ 28
Paint, felt, hardware, etc.	£ 26
Timber	£ 62
Total Cost of Basic Collector	£276
Cost per square metre	£ 16.24

Cost of Auxiliary Equipment

Pump	£ 43
Copper pipe (28mm), fittings, side tank, etc.	£ 36
Pre-heat tank (730l fibre glass tank)	£ 80
Insulation board for tank	£ 10
Electronic and electrical components	£ 30
Total cost of auxiliary equipment	£199

RESULTS ACHIEVED

A comprehensive monitoring of the solar water heating was not possible financially. However, the evidence which has been gathered indicates that the collector is utilizing about a half of the available solar energy and is saving in the region of 70% of the electricity previously used. The performance is thus better than anticipated, and the Iona Solar Water Heater should pay for itself within the five year period. A similar system may thus save each season about 5390 kWh, or 317 kWh per square metre of collector surface, of electricity. It must be stressed that the actual saving depends on the use of hot water as well as the size and design of the solar water heater.

However, there have also been disappointments. In particular thermal stratification in the pre-heat tank (that is, the hottest water remaining at the top) has been disturbed whenever the pump is switched on. Although improved baffles might help, it is not proposed to experiment further with them since this problem will not occur if an 'indirect' system is introduced.

More serious has been frost damage. The controller has a frost-stat to protect the system against frost damage, but when the Centre was empty during the winters of 78/79 and 79/80 the system was drained and the controller switched off. Unfortunately, draining the system was not wholly successful and frost damaged sections of the 15mm copper pipes. While repairs were being carried out in the Spring 1980, several sheets of glass, taken off the collector to facilitate repairs, were smashed by high winds. It took a surprising length of time to get the correct glass to Iona.

The decision has since been taken to convert to an indirect system, and this work is soon to be carried out. It is intended to use an old copper cylinder as a heat exchanger within the pre-heat tank. This will enable antifreeze to be added to the solar circuit. The change to indirect will also overcome the difficulty which was experienced in maintaining thermal stratification.

CONCLUSION

Although the Iona Community were able to draw on goodwill and expertise, the actual building work was not difficult and was carried out by a group of enthusiasts whose occupations ranged from Minister to Funeral Undertaker. Do-it-yourself Solar Energy is feasible and can be economic.

The Youth Centre at Iona was well suited for the application of solar water heating, particularly since hot water is required only during the season and is used mainly at mealtimes during day light hours. There are no doubt many other buildings in the highlands and islands with similar requirements.

It is hoped that the experience gained on Iona will encourage others to make use of solar energy.

ACKNOWLEDGEMENT

I would like to express my gratitude to the Iona Community and the Church of Scotland SRT Project for encouragement, to Kerr MacGregor and others for expert advice, to the companies who gifted material, and to the members of my congregation who carried out the work.

REFERENCES

Carrie, J.G. (1978), The Iona Solar Energy Project, Church of Scotland SRT Project, Edinburgh.

MacGregor, A.W.K. (1978), Proc. ISES International Congress, New Delhi.

Traditional Houses and Energy Use in Barra and South Uist

Peter Whyman

ABSTRACT

The Hebridean black house had at its centre an open peat fire which burned continuously. The traditional "Department plan" three apartment house was designed for continuous heating by two open fireplaces, one at either gable.

The difficulty of obtaining adequate supplies of peat and the advent of electricity have brought about the introduction of an intermittent heating cycle which is costly and inefficient in modifying the damp interior climate.

Re-introduction of continuous heating must depend on two factors: the cheapest possible fuel and the most economical use of fuel.

Simple mechanical aids to the extraction and drying of peat could make the use of this fuel in improved continuous burning appliances possible at low cost.

It is suggested that the black house was more efficient in its use of energy than the "Department plan" house and that given a clean source of heat it offers a better model for development.

KEYWORDS

Black house. "Department plan" house. Continuous heating. Peat extraction and use as fuel.

The observations in this paper arose from a survey of houses on Barra and South Uist in 1972 to 1974. The survey was commissioned by the Highlands and Islands Development Board, (Whyman 1972) to encourage the islanders to take up improvement grants and to bring their house up to tolerable standard as defined in the Housing (Scotland) Act 1969. A survey by the Inverness shire County Council had disclosed that the percentage of sub-standard houses was over 50% (HIDB 1974) but the number of applications for improvement grants was very low indeed.

Thirty four houses on Barra and a similar number on South Uist were surveyed in detail as a basis of improvement grant applications. Many more were visited but not surveyed in detail. All had mains electricity although by no means all had a piped water supply within the house. Although it was not the main subject of the survey it was noticed that a radical change in the mode of space heating had come about: electricity was in favour both for cooking and for space heating and in some cases the solid fuel stoves, ranges and fireplaces had been taken out.

The climatic data of the outer islands is well known. It is relatively mild, very windy - on Barra the wind is at force 8 or more on average of one day in ten - and wet. The mild temperature means that although frost is comparatively rare in the winter, real warmth is also comparatively rare in summer and this combined with the humidity means that comfortable conditions cannot be sustainted indoors without artificial heating throughout the year.

The ancient house form, the "black house", originated before Viking times and persisted into the present century. It was of oval shape with massive walls some 1.8 m thick, with a roof of turf covered with thatch. In the centre was a fireplace on which peat was burned. There was no chimney: the smoke escaped through the thatch. (The reason for the name is obvious). None are now inhabited on Barra or South Uist and such as still exist are in ruins but a fine example has been restored at Arnol in Lewis by DOE (Fenton 1978).

The commonest inhabited house form is the "Department" house, built from standard plans produced by the Department of Agriculture and its successors from the end of the first world war. The plans varied in dimensional details from time to time but the essential feature of the "Department" house is that it is a rectangular three apartment house with a fireplace at each end. (Fig. 1)
The walls are of masonry, 600 mm thick and the roof is slate, laid directly on boarding supported by rafters. Internally the house is lined with matchboarding or plaster on lathing, the same materials being used on timber framing to make the partitions. A scullery, or additional room, is often provided as a lean-to extension to the back of the house. This is covered by a slate roof of lower pitch than the main house and although it looks like an addition it is an original feature of some of the inter-war standard plans.

"Department" houses often have rooms in the roof space, lit by skylights or dormer windows.

The heating and cooking arrangements are always the same: there is a chimney at each gable. One serves the kitchen and the other is for space heating the opposite apartment which is used variously as a living room, a parlour or a bedroom. The kitchen was originally equipped with an iron open grate, later replaced with an iron range such as the "Modern Mistress" and latterly with an enclosed stove such as the "Rayburn". All burned peat and the kitchen fire at least was continuous-burning. It is now commonplace to find a two or three bar radiant electric heater (with imitation coal effect) in the living room or parlour and an electric cooker in the kitchen. Continuous-burning solid fuel has been abandoned in favour of clean, convenient, (but very expensive) electricity. The change in the heating cycle is radical, instead of the gentle but continuous output of radiant heat warming up the surfaces of the walls we have short bursts of intense radiant heat. Ventilation is

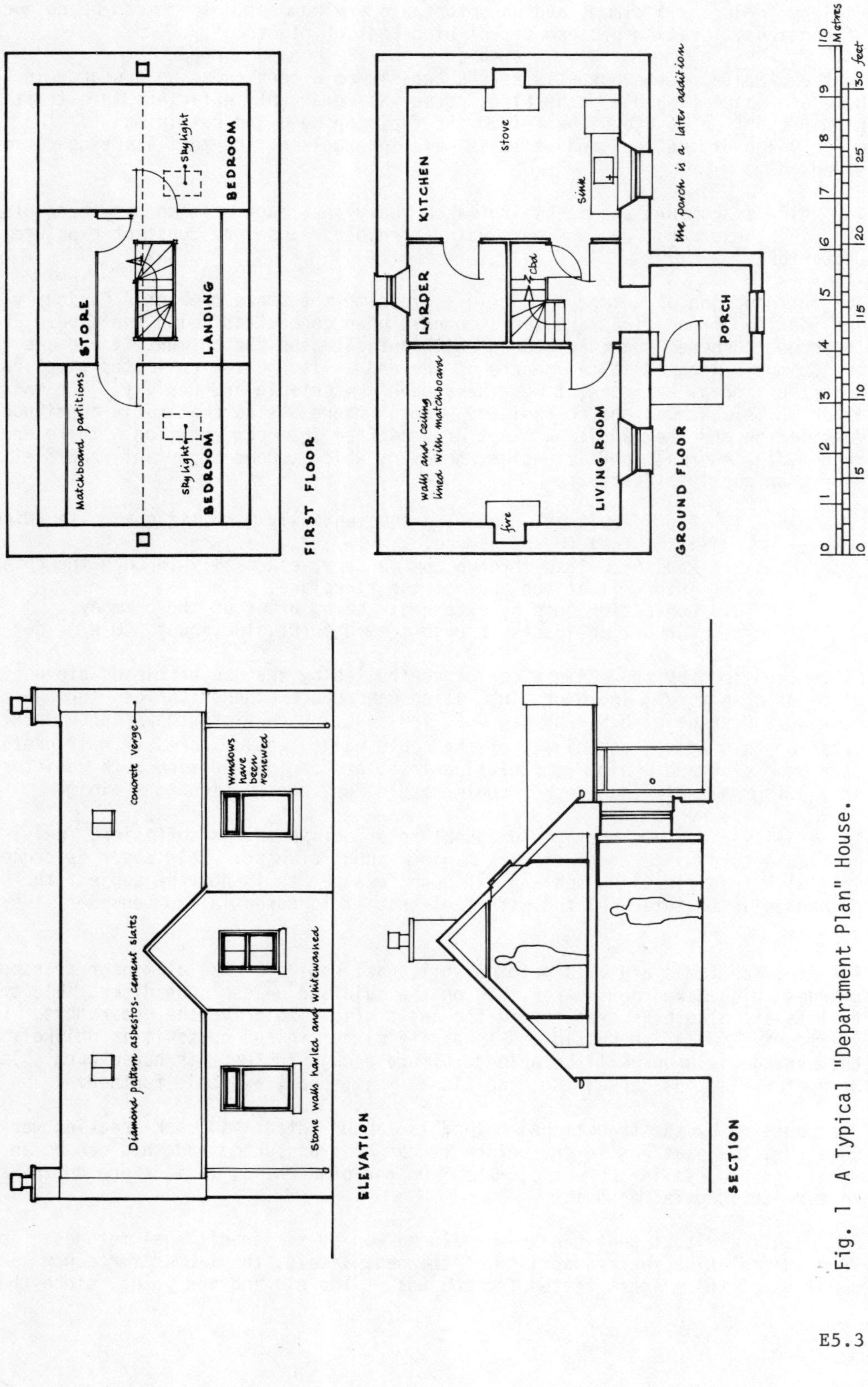

Fig. 1 A Typical "Department Plan" House.

at a minimum since the flues are closed and the masonry becomes damp by condensation, producing a clammy and uncomfortable environment. There can be no question of continuous heating because of the high cost of electricity.

In the climate of the outer isles the benefit of a continuous heat source in the home are unquestionable. The black house was remarkably efficient in extracting a high proportion of useful heat from the fuel consumed and retaining it. The insulation value of the roof and walls is well in excess of the most exacting current standards.

The price in comfort and in health was a heavy one; many contemporary accounts describe the chronic eye and respiratory troubles caused by constant exposure to peat smoke.

The introduction of chimneys and flues overcame the smoke problem. Curiously, duplication of the fireplace seems to have been concomitant with the flue. Even in the round-cornered thatched houses which antedate the "Department" there are always two chimneys located at the centre of the end walls. Given unlimited supplies of peat the thermal environment must have been comfortable and healthy. The massive outer walls absorbed and re-radiated warmth, there was automatically continuous ventilation and the fabric was kept dry and free from condensation. There are, fundemental thermal inefficiencies, however, which become important when fuel is scarce or costly. These are:

1. Two fireplaces, requiring substantially more (if not quite twice as much) fuel to run as one.
2. Radiation loss through the masonry, especially through the relatively thin wall at the back of the fireplace.
3. Convection loss by excess air being drawn up the chimney.
4. Conduction losses through the roof (U value about 2.0 W/m^2 deg C)

Convection loss by the chimney can be controlled by the use of closed stoves in place of open fires, and added insulation can reduce conduction loss through the roof to a U value of 0.5 W/m^2 deg C at low cost, but what further can be done? Exterior insulation, especially of the gable walls, would increase their usefulness as a heat sink and would cut radiation loss, and central heating with radiators driven from a solid fuel cooker could reduce fuel consumption to a minimum.

These palliatives are costly and cannot be effective unless sufficient fuel is available to provide the necessary minimum input of heat. This paper is concerned only with traditional houses and although fuel per se is not the subject the design of houses is so inter-related with fuel that it is essential to consider it briefly here.

All imported fuels are at a premium: oil, gas, coal and coke all incur transport charges which make them dearer than on the mainland unless subsidised. Electricity tariffs are also high, reflecting the costs of oil to drive the generators. Even if imported fuels are subsidised to parity with mainland costs it is unlikely that the average islander will be able to afford enough to keep his house warm. It may be worthwhile considering the traditional fuel; peat, a little further.

Peat cutting by the traditional method is labour-intensive, back-breaking work. Once cut, the peat has to dry before it can be transported and this can be an uncertain process in the wet climate. On the other hand, it is there on the spot, in very substantial deposits

It is unlikely that peat can ever again be won by the traditional methods. Apart from the relative inaccessability of the peat itself, the labour force now largely consists of those least fitted for the work - the old and the young, since the able-bodied tend to emigrate for their working lives to the mainland or to the merchant service. The best hope lies in less labour-intensive peat winning. In an unpublished undergraduate thesis (Provan, 1975) it was pointed out that although machine extraction of peat on a large scale for industry had been developed, little had been done on a domestic scale and it was suggested that farm machinery could fairly easily be adapted to cut and break up the surface for initial drying. Once dried to about 50% moisture content the peat could be scooped up and either pressed into briguettes at about 30% moisture content or dried as pulverised fuel in machinery already developed by the National Coal Board for grass drying, but adapted to burn pulverised peat instead of pulverised coal. The energy for driving all the machinery and for the drying process could be derived from peat economically.

Perhaps the traditional house and the traditional source of energy on the Islands is not dead, only dormant. Given the application of relatively low key technology the problems of peat winning could be overcome. Once won, peat must be used with the utmost economy since it is non-renewable. A house plan having a heavily insulated, thermally massive shell, with the heat source at the heart, drived from the principle of the black house more than the "Department" house, seems an obvious model for the future.

REFERENCES

Fenton, Alexander (1978) The Island Blackhouse, HMSO

Highlands and Islands Development Board (1974) Rural Housing in the Highlands and Islands. Occasional Bulletin 5

Inverness County Council (1971) Improvement of Rural Housing in the Outer Hebrides

Provan, Archie (1975) A Hebridean Cooperative. Unpublished undergraduate thesis. Department of Architecture and Building Science, University of Strathclyde

Whyman, Peter (1974) Housing Improvement Surveys, Barra, 1972 HIBD Special Report 12.

A Low Energy House, North Ronaldsay, Orkney

Robin Jacques

Department of Building, UMIST, Manchester, UK

ABSTRACT

The preliminary design for a private dwelling house is described. The brief required that the feasibility of 'alternative' energy sources be explored, especially wind power. It was found that feasibility depended upon a building design which permits the minimisation and control of energy demand for space heating. The design described is calculated to have a maximum rate of heat loss of 3.05 KW for a 22°C temperature difference, and a seasonal heat requirement of 16 GJ. The wind regime of the island, the long cool summer days and mild winters, make it practical to utilise wind energy augmented by passive solar collection. A heat store of 6,200 litres of water is incorporated. Mechanical ventilation with heat recovery and insulating shutters are included. It is concluded that, in general, the energy conscious design of buildings and their services systems is the best answer to an uncertain energy future.

KEYWORDS

Low energy house; wind power; passive solar energy; high insulation; heat recovery; energy conservation.

INTRODUCTION

This paper will describe the design for a private dwelling on the island of North Ronaldsay, Orkney. Its special interest derives from the client's brief for a house independent of imported fuel supplies using, if practicable, wind energy, and from the special circumstances of its remote island site.

The paper is written at an intermediate stage in the design process and not all considerations are complete nor all decisions made. It was felt however that the thinking involved up to the present stage might be of interest to those concerned with energy problems and energy efficiency.

THE SITE

North Ronaldsay is the northernmost inhabited island of the Orkneys, with a population of 126 people. It is almost entirely flat and low lying. It is swept by some of the most persistent winds in the British Isles, which prevent the growth of any substantial trees and only allow vegetables, shrubs etc to thrive in the lee of walls or buildings. The winters are surprisingly mild, but some space heating is required in summer. Table 1 shows the average day maximum and average night minimum temperatures for the period 1947-1978 (^{o}C).

TABLE 1 MONTHLY AVERAGE TEMPERATURES, 1947-1978

Month	Average Max	Day Min	Month	Average Max	Day Min
Jan	5.5	2.1	July	15.1	9.5
Feb	5.5	1.5	August	15.3	9.8
Mar	7.0	2.4	September	13.7	8.7
April	9.1	3.2	October	11.4	7.1
May	11.4	5.6	November	7.9	4.2
June	13.8	7.8	December	6.2	2.8

Source; Kirkwall Meteorological Office

Reference to the Meteorological Office Summary for 1977 (HMSO 1979) shows that at Kirkwall Airport wind strength was 5.67 m/s (11 knots) and over for 5,247 hours (out of a possible 8,760), and was over 11.34 m/s (22 knots) for 1,108 hours of those. BRE Digest 210 (HMSO 1978) shows an hourly mean wind speed exceeded 50% of the time for the period 1965-1973 of about 6.5 m/s at a height of 10 metres.

Inspection of detailed wind records for 1979 taken at three hourly intervals from 0900-2100 each day, shows that there were rarely more than two consecutive days of consistently low wind speeds. (Barbour 1980). April, June, July and August each had one period of 5 or 6 consecutive days when the average wind speed was only 3 m/s, with a maximum during those periods of 6-8 m/s. February had one such period of 3 consecutive days, but apart from this all periods of such low winds lasted a maximum of 2 days.

There are no mains energy supplies or remaining indigenous fossil fuels on the island. Coal is imported for space and domestic hot water heating, oil for individual domestic diesel electricity generation, and bottle gas for cooking. Controversy surrounds the possible future provision of a mains electricity supply to the island. Uranium has been discovered in economic concentrations but the Island's Council is opposed to its exploitation.

It is in this context that Dr and Mrs Woodbridge propose to develop a derelict croft as a family home. The site is about 10 metres above sea level on a promontory facing the open sea to the south and west. To the east the view stretches along the coast, whilst to the north lies a fresh water lake, Loch Gretchen, which is important to the client's ornithological interests. Access to the site is from the north, where the existing long house and its outbuildings are located roughly at the apex of a triangular site, the base of which is the southern coastline at a distance of about 200 metres from the house.

The existing buildings consist of a traditional long house with an extension to the west end of the south face, a byre in reasonable condition, and various other outbuildings in an advanced state of dilapidation. The house has been unoccupied for about twenty years but the basic structure is sound. The wall construction is of local masonry some 600 mm thick, bedded to drain outwards, in a sea sand mortar. The walls have a high moisture content and are unlikely to dry out due to the climate, the cumulative infusion of airborne salts and the absence of a damp proof course.

SOME POINTS FROM THE FEASIBILITY STUDY

Part of our study included the exploration of all possible sources and technologies of energy supply, conversion and storage, including methane, hydrogen and total energy packages such as the Fiat 'Totem' system. Many lay outside the scope of individuals to apply at the domestic level, others are at too early a stage of development to be adopted with confidence. In virtually all cases the initial costs to an individual for domestic usage are prohibitive by normal accounting assumptions. It was concluded however that it was practical to utilise wind energy, augmented by passive solar collection, with a stand-by diesel generator.

The above observations apply equally to wind generators. Reliability and durability present problems. Energy switching and control systems are in the early stages of development, energy storage is difficult and costly, the capital costs of the installations are high, and little evidence exists to back up the performance claims of many of the manufacturers.

On the other hand the wind regime of the island is very favourable, and wind uniquely among the ambient sources tends to deliver most energy when it is most needed. Additionally there are no mains supplies to compete in the economic equation, and wind generators are capable of supplying electrical power as well as heat.

The adoption of passive solar collection is particularly appropriate to this situation. The sun is the only ambient source which can make a positive energy contribution to buildings without the benefit of an installation (Burberry 1978). To capitalise on this requires specific attention to various aspects of building design, of which more follows. The special contribution of solar gain to this site is however due to the very long but cool summer days. Thus there is a need for a space heating input during the months when the sun is best able to provide it. MacGregor (undated) concludes that annual useful gain at Lerwick (60°N) can be about 35% higher than at Kew (53°N) due to the higher heating loads in the North, particularly outside the winter months. The possibility of incorporating a conservatory for solar gain is attractive due to its other benefits on such an exposed site, which include horticulture and its use as an extension of the living space.

Heat storage was considered best achieved by the use of water, since this could be used also as the heat distribution medium in a more or less conventional hot water radiator system. Some heat storage is anticipated in the walls and floor of the conservatory. Electrical storage is expected to be based upon ex GPO exchange batteries.

Clearly, the size of wind generator needed and the need for thermal storage to cover periods of no wind and no sun can be minimised by adopting an energy conserving design for the dwelling. The greatest potential for achieving reductions from current norms of domestic energy consumption lies in space heating, and to a smaller extent in domestic water heating. The energy requirements for electrical appliances are generally irreducible, since to do so would interfere with matters such as health, hygiene, communications, information and entertainment. Some reduction in peak demand may be achieved by behavioural adaptations, perhaps aided by electronic control systems which avoid the simultaneous use of equipment of high demand. Table 2 shows figures for the average annual domestic energy consumption of a UK household (BRE Digest 191).

TABLE 2 AVERAGE ANNUAL DOMESTIC ENERGY CONSUMTPIONS (GJ)

	Net	Gross	Useful (approx)
Space heating	52	74	30
Water heating	17	30	9
Cooking	7	16	5
Lighting, TV etc.	5	18	5
	81	138	49

Reduction of Space Heating Demand

Building shape, orientation, layout, fenestration, insulation and ventilation are primarily instrumental in determining the energy demand for space heating. It is not possible to separate these entirely, decisions in some areas having an interdependence with the options in others.

Various experimental houses have shown that it is possible to make major reductions in space heating demand. The Saskatchewan Conservation House, operating in conditions having a design temperature difference of 55°C, has a space heating requirement of only 5.1 GJ per annum. (Besant, Dumont and Schoenau 1979). Many other examples exist, such as the Salford houses (Randell 1979), Granada TV's'House for the Future' (Stephen 1980), the BRE experimental houses (AJ 1980), those at the Centre for Alternative Technology (Todd 1978), and numerous other 'Landmark houses' (Littler 1980) which show that the annual heating demand can be reduced dramatically through attention to the various facets of building design.

Also of particular importance in the consideration of the energy supply technology is the reduction of the peak energy demand. In the Salford houses with a mean outside temperature of about -4°C at the turn of the year, the temperature of the living room was maintained at about 21.5°C, with the coolest room at about 17.5°C, by a heat input of 2.5 kW (Randell 1979). Steinmuller and Bruno (1978) cite figures of 16 kW for a Normal House, 6 kW for the Swedish Standards House, and 2 kW for the Philips Experimental Standards House in Hamburg.

DESCRIPTION OF THE DESIGN

The design seeks to achieve a low outline consistent with minimum surface area for the volume enclosed. The form is in part dictated by the client's desire to use constructional methods and materials suited to local island builders. The fenestration aims for an advantageous balance of solar gain and heat loss consistent with the needs for view. The layout of rooms within the plan is influenced by the desire to keep the piercing of walls for natural light and view to a minimum on northerly elevations. This is obviously facilitated by the fact that the open sea views are southerly, though the need for spotting bird movements on Loch Gretchen necessitates some windows on the north side. Doorways are protected either by the conservatory or by the glazed link with the long house. (See Figs. 1 and 2).

Wall construction is proposed as an inner skin of dense concrete blockwork with plaster finish, 150 mm polystyrene slab insulation, 25 mm cavity, and an external skin of concrete blockwork rendered and painted. This gives a U value of 0.165. It provides some internal mass for a slow thermal response time which was considered preferable for comfort. A solid internal skin also permits easy traditional uses such as hanging bookshelves and withstanding high velocity children.

The roof has a 35° pitch and is finished with asbestos-cement slates for the

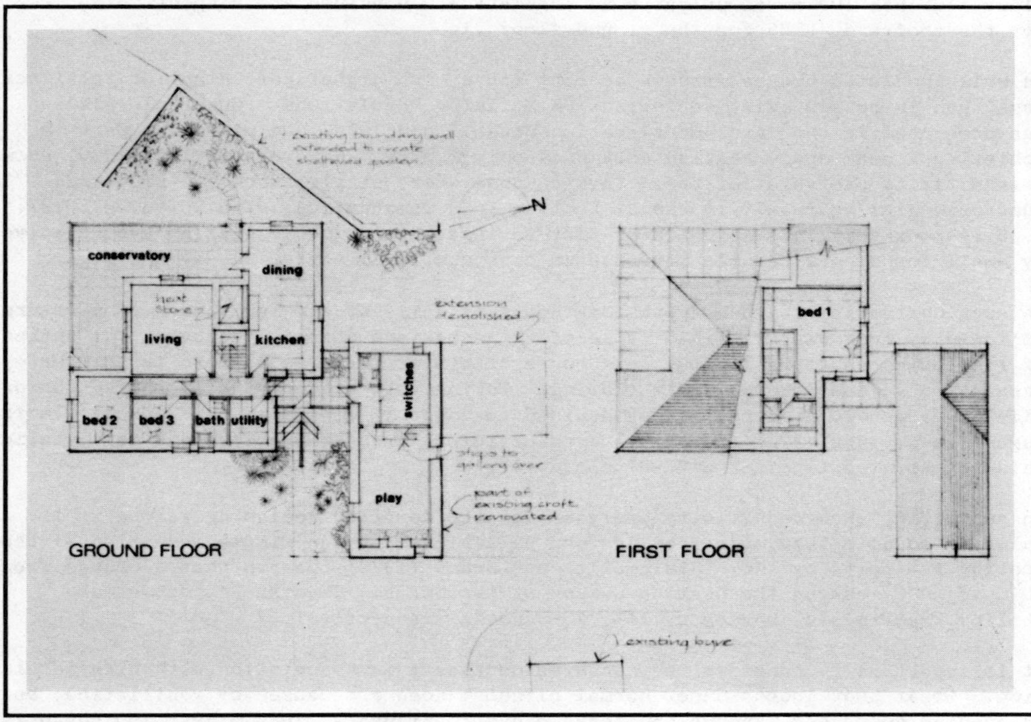

Fig. 1. Plans.

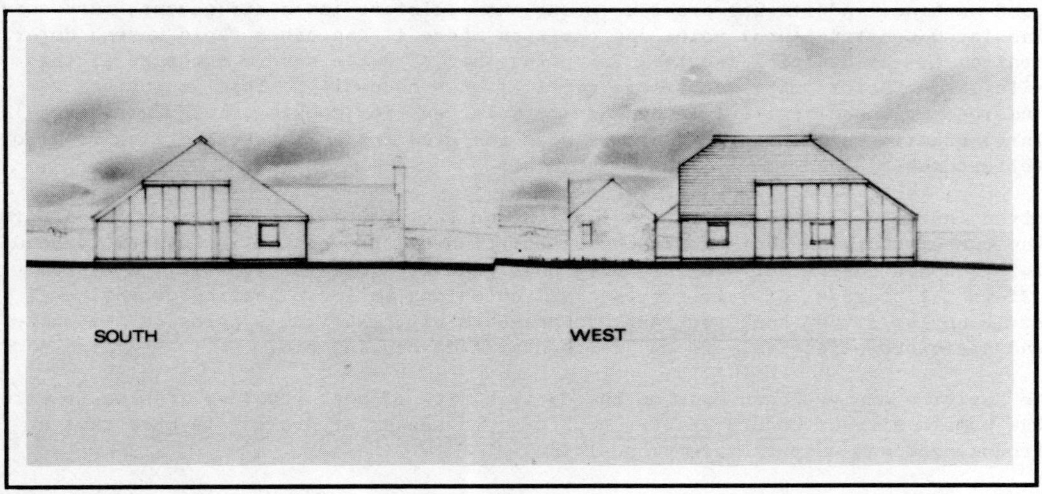

Fig. 2. Elevations.

extreme exposure, incorporating 150 mm of glass fibre insulation. The concrete floor slab has 100 mm of polystyrene insulation. U values are respectively .202 for the roof and .225 for the ground floor slab.

In well insulated houses windows account for a much higher percentage of total heat loss than in houses built to current UK Building Regulations. Quadruple glazing has been used in the Wates Conservation House (Todd 1978) and plays its part in achieving a peak space heating demand of only 1.2 kW. However Littler (1980) shows a sensitivity analysis for the Autarkic House where single, double, triple and quadruple glazing result in annual heating load demands (GJ) of 5.29, 3.63, 3.27, 3.06 respectively. This indicates that savings beyond double glazing are relatively small for a considerable increase in constructional costs.

In such northerly latitudes with long hours of daylight and low summer temperatures, followed by long winter nights, some of the equations change. However, the option of insulating shutters for the long hours of winter darkness achieve far higher insulation values than multiple glazing. Taking into account the increased comfort offered by double glazing, it was decided to adopt 12 mm sealed unit double glazing augmented by insulating shutters. With efficient draught-proofing and edge sealing this is expected to offer a U value of .57.

In conducting such sensitivity analyses for the detailed design decisions it is salutary to note that major savings can be achieved by the simple reduction of the heating set point by $1^{\circ}C$. Steinmuller and Bruno (1979) observe that a change from $20^{\circ}C$ to $19^{\circ}C$ reduces the heating demand of the Normal, Swedish Standards, and Philips Experimental houses by 11%, 13% and 24% respectively.

It is important to consider the measures necessary in association with high insulation. Great heat losses under normal circumstances are caused by ventilation, both in opening doors and windows and leakage due to draughts. BRE Digest 190 suggests that 21 GJ out of a total of 76 GJ of seasonal heating demand would be due to ventilation losses for a typical semi-detached house of 100 m^2. Littler (1980) suggests that in a well insulated house ventilation accounts for about 40% of the space heating demand. Thus controlled ventilation takes on a far greater importance in a well insulated house, both for this reason and for the avoidance of condensation. Additional problems occur when moisture laden air escapes into non-insulated constructional voids and cavities where it can cause rapid unseen deterioration of the building fabric. Taken together with the severe exposure of the site these factors make mechanical ventilation a necessity. This is further reinforced by the client's intended use of Calor gas for cooking, which generates large quantities of water vapour, as does the need for an internal drying room for wet outdoor clothes etc.

Given the need for mechanical ventilation and the proposal to utilise warm air from the conservatory in the summer, the elaboration of the ventilation system to handle this becomes desirable. Additionally heat recovery from the ventilation system offers the prospect of very substantial reductions in space heating demand. Efficiencies of 70% heat recovery are possible with existing systems on the market, and therefore effectively re-use waste heat from cooking etc.

No decision has yet been made on the desirability of heat recovery from waste water. The possibility of odours, filter problems etc appear at present to make this a troublesome and unattractive proposition.

Performance Predictions

The measures outlined above are calculated to result in a maximum rate of heat loss of 3.05 kW for a temperature difference of 22 degrees C for the house as a whole.

Using the degree day method described by Burberry (1979b)a seasonal heat requirement of 16 GJ is predicted. However, cross checking with different methods of calculation produces some variation in this figure in both directions, and it may be as high as 20 GJ.

Preliminary calculations based on the monthly temperature averages suggest that, in July and August, casual gains will nearly balance heat loss. Either side of these months space heating input will be required. The conservatory is generally expected to meet space heating needs from June to September, though its contribution could extend beyond these months in some seasons. It has not yet been decided whether it is worthwhile double glazing the conservatory. Whilst for single glazing total annual heat losses far exceed heat gains in the conservatory, the critical factor is the seasonal advantage. During the period when the solar contribution of the conservatory is expected, gains will exceed losses. It is further felt that double glazing will be likely to preclude the conservatory's use by plants and humans due to excessive air temperatures.

The major input to the space heating will come via the main heat store powered by resistance heating direct from the wind generator. The heat store consists of 6,200 litres of water surrounded by 150 mm of insulation. It is located internally between the main living areas for distribution efficiency. Working on a temperature range of 40°C-70°C this has a capacity of 220 kWh, sufficient for three days of space heating during a period of no wind at the continuous maximum rate of demand of 3.05 kW. Only rarely will such a period of demand occur and the heat store will more normally provide for a considerably longer period, depending on ambient outside air temperature.

Generally the performance calculations have been based on the IHVE Guide (1970) and Burberry's (1979a) worksheets which have proved to be a useful aid. Later work will need to use more sophisticated models which include dynamic thermal performance, the effects and contribution of insolation, and the energy input patterns in relation to heating and domestic electricity load patterns.

CONCLUDING REMARKS

It has been difficult to conclude that the current state of the art in 'alternative' energy technology has much to offer at the individual domestic level. Proposals to tap wind or other sources of energy at a larger scale would appear far more promising since they offer levels of convenience similar to those pertaining generally today. Electricity is the most important form of energy supply and the most difficult to provide through renewable sources. More options exist for space heating, yet even in this area the need for storage constitutes an expense which has been unnecessary in recent decades. It appears far more convenient and cheaper to store fuel rather than energy, and developments such as hydrogen power with a broad spectrum of uses look promising.

Once levels of energy demand have been attained which make reliance upon ambient sources practical, the demand on mains supplies is similarly greatly reduced. Great economies are then possible without the need to invest capital in individual energy conversion and storage devices. It is the unusual factors of <u>no</u> mains supplies and a suitable wind regime which makes the adoption of a wind generator a feasible proposition, a circumstance not uncommon in remote communities. Additionally, the very severe exposure which makes a wind generator possible also imposes rigours and disciplines upon those who live in remote communities which are less familiar to the majority of urban dwellers. For example, the clients for this project are already used to the behavioural adaptations involved in living with a diesel generator, which are similar to those required by a wind generator, though less extreme.

The one constant which can be identified in the absence of a new, ubiquitous and cheap energy source is that good energy conscious design of buildings and their services systems is the best insurance against an uncertain energy future available to us at present.

ACKNOWLEDGEMENT

The project described has been carried out in association with J. D. Adams.

REFERENCES

The Architects Journal. Houses in the 80s, Energy Systems. 16 Jan 1980. pp. 143-147.

Barbour, D. (1980). Personal communication. Department of Applied Physics, University of Strathclyde.

Besant, R.W., R.S. Dumont, and G. Schoenau (1979). The Saskatchewan Conservation House: Some Preliminary Results. Energy and Buildings, 2. pp. 163-174.

Building Research Establishment. Heat Losses from Dwellings. BRE Digest 190. HMSO, June 1976.

Building Research Establishment. Energy Consumption and Conservation in Buildings. BRE Digest 191. HMSO, July 1976.

Building Research Establishment. Principles of Natural Ventilation. BRE Digest 210. HMSO, February 1978.

Burberry, P. (1978). Ambient Energy - Criteria for Building Design. In Randell, J.E. (Ed), Ambient Energy and Building Design, The Construction Press, London. Chap.12, pp. 143-162.

Burberry, P. (1979a). Predictions of Thermal Performance-1. The Architect's Journal, 24 Oct 1979, pp. 893-905.

Burberry, P. (1979b). Predictions of Thermal Performance-3. The Architect's Journal, 7 Nov 1979, pp. 993-1004.

IHVE Guide, (1970). Inst. Heating and Ventilating Engrs, 1971. Book A.

Littler, J.G.F. (1980). Low Energy Houses. Ambient Energy. Vol. 1, No. 1. Jan 1980, pp. 57-62.

MacGregor, A.W.K. (Undated). The Potential for Solar Assisted Space Heating at High Latitudes in the British Isles. Department of Mechanical Engineering, Napier College, Edinburgh

Meteorological Office. Monthly Weather Report, Summary 1977, Vol. 94, No. 13, HMSO 1979.

Randell, J.E. (1979). Low energy housing - Salford's answer. Building Services & Environmental Engineer, June 1979. pp. 18-21.

Steinmuller, B., and R Bruno (1979). The Energy Requirements of Buildings. Energy and Buildings, 2. pp. 225-235.

Stephen, F.R. (1980). Performance Monitoring of Low Energy House, Macclesfield. Ambient Energy. Vol. 1, No. 1, Jan 1980. pp.29-46.

Todd, R.W. (1978). Low Energy Housing at the National Centre for Alternative Technology. In Randell, J.E. (Ed). Ambient Energy and Building Design. The Construction Press, London. Chap. 7. pp. 75-89.

Low Energy Project—First School at Tidcombe Lane, Tiverton for Devon County Council

J. M. Collins, M. E. P. Driver, J. A. Gill, R. S. Knight, D. Moore, R. C. Peachey, I. J. Phillpotts, D. J. J. Sutherland and R. B. Vickery

Devon County Architect's Department, Broadwalk House, Southernhay West, Exeter, Devon EX1 1TX, UK

ABSTRACT

Report on a low energy school which sets out to demonstrate the conservation of natural resources in its design and educational use; it includes descriptions of the solar installation, kitchen waste heat recovery unit and rainwater recycling powered by wind pump.

KEYWORDS

Energy Conservation; Primary Schools; Flat-plate Solar Collectors; Solar Water Heating; Solar Space Heating; Waste Heat Recovery; Windpumps; Education.

INTRODUCTION

Architects work in a time perspective of -2000 to +100 years. They play the role of art and economic historian and search for the link between artefact and its context in an effort to employ meaningful symbols in today's design tasks. They are absorbed in matching the qualities of materials with the anticipated life of the object. They are problem solvers as well as problem seekers and in that way have been concerned with the question of energy for many years.

The concern of the County Architect in Devon led, among other things, to the establishment in 1975 of a permanent Energy Working Party which maintains a broad view of future strategies in the County. In 1976, after discussion with the Education Officer, the proposal to set aside a school project for special research into low energy solutions was approved by Committee. The Tiverton School lay far enough ahead in the programme to allow time for development studies to be carried out by both architect and client.

The project was viewed not as a prototype, which would have imposed additional limitations, but as a unique exercise in exploring the relationship to be sought between the design and educational use of a school which both sets out to be conservative in the use of natural resources and to demonstrate the virtues of conservation. It will act as a public demonstration of the authority's concern to guide and teach others both within and outside the education service in the work of conserving natural resources.

DEVELOPMENT STUDIES

Site, energy and building studies proceeded concurrently and were deliberately complementary and interrelated. The design team adopted a practical-realist approach which explored but did not necessarily exploit the many and various conservation techniques. Working objectives were to demonstrate the validity of a building design based on energy conservation principles, to explore the maximum extent to which a School can reduce energy and resources intake and achieve autonomy, and to ensure that the building solution makes a low demand on resources for the upkeep of the fabric during its life. The temptation to expand floor area and force the architect to use poor quality materials has to be resisted.

Site Study

The site, of 0.93 hectares, is long and somewhat narrow from east to west. It is bounded by rising ground to north and south, tends to hollow towards the centre and falls 6m from Tidcombe Lane on the east to the Ailsa Brook on the west. On the far side of Tidcombe Lane lies the Grand Western Canal contained within a 6 m high embankment. Drainage, water and electric services are readily available; gas is at a distance and would involve a considerable connection charge. To prevent flooding of the Ailsa Brook the S.W.W.A. require provision to be made for holding on site the whole of the run off with controlled outflow to the brook. An open ditch has been arranged on the south boundary to collect this storm water which then discharges into a pond with restricted outflow at its bottom end. Advantage is taken to create here a wetland habitat for the children's study.

Studies were carried out into the use of the site's natural assets to serve the building. There was investigation into the extraction of heat from the canal water by means of a heat pump and,while the sheltered nature of the site appeared to mitigate against the use of wind power, both these proposals were considered in the energy studies. After the decision was taken to use solar energy for heating and hot water, observation was made of any shading of the building area. The only shadows came from the three fully grown deciduous oak trees near the entrance. Exercises with a shadow angle protractor to produce sun path diagrams indicated such shading confined to the lowest panels during very short periods during the latter part of the day in winter when radiation is small.

Building Study

It was established at an early stage that the building envelope should provide the major climate modifier and require the minimum of purchased energy to maintain a satisfactory internal environment. In terms of building shape and form, such a requirement is best met by simple massing and by enclosing a maximum volume in a minimum perimeter. The single access to the site called for the siting of the building to be located in the narrow neck at the eastern end. Here a single storey school would of necessity be somewhat elongated. It was therefore proposed to plan the school on two levels - the upper floor being educationally and aesthetically acceptable. The absorber panels for a solar heating system can also be satisfactorily included on this basic structure and in turn dictate the building's orientation. Preliminary energy studies favoured a heavy mass construction for good thermal control and a loadbearing structure supporting pitched roofs of comparatively small economic spans has been adopted. This construction method also made economic sense.

Fig. 1 model of the project

A major factor among these causing loss of heat from a building is the leakage of air through windows, doors and other openings in the external envelope. However, a completely sealed school was recognised as having fairly substantial disadvantages, particularly perhaps in Devon in the summer. The energy argument also lends support to the decision to opt for the privilege of controlled use of external openings rather than a sealed envelope. During the heating season entry will only be through draught lobbies and ventilation through nite-air hit-and-miss fittings at the heads of windows, the air being drawn out through extract fans over the toilet areas. The main entrance for children is in fact a sheltered but unheated area of translucent material which can accommodate a full class and may also be used for teaching purposes when the climate permits.

It was agreed that the external walling and roof were to be insulated to give a max U value of 0.5 W $m^2/^oC$. The proportion of heat loss through the walling compared with that through the glazed areas of external skin is small and to assist the overall performance glazing will be restricted to 25% of the total area of external skin and double glazed. This ratio of double glazed window/wall gives an overall U value for the external skin of 1.19 W $m^2/^oC$. The reduction in window area also modifies the effect of solar gain in summer months while producing more interesting view windows from the interior.

The reduction in glazing area on the perimeter has necessitated a careful look at the daylighting of teaching areas. Six rooflights have been introduced over deep plan areas.

The formula: Average df = $\frac{T A_3 \theta}{2 \Sigma A (1-P)}$ has been used to establish that the 5% <u>average</u> daylight factor (recommended by 1977 I.E.S. Code) has been achieved over the major part of the teaching areas. Spot checks using BRE daylight protractors (overcast sky) indicate that a <u>minimum</u> of 2% daylight factor has also largely been achieved by careful placement of windows. The use of artificial lighting can therefore be reduced to a minimum during normal school hours.

The school is largely open plan. The question of insulation against noise is confined to two main areas. The brief asked for the assembly hall to be closed off from the rest of the school. The hall, also used for dining, has, with the kitchen, been placed in the north east corner of the building and is approached only through two pairs of double doors. The decision to use a loadbearing structure has determined a wall thickness of 225 mm between these areas and the rest of the school giving an insulation value of 55 db. Home bases are sheltered by enclosure on three sides.

Energy Study

In their energy study the in-house engineers were assisted by Specialist consultants and members of the Physics and Chemical Engineering departments of Exeter University.[1] The explicit requirement of the energy consuming environmental engineering services was to provide an adequate and satisfactory level of visual and thermal comfort within the occupied building envelope but with the reduced use of finite or supplied energy resources. The conservation principle was to be further extended to the provision of hot water for domestic use and energy used in the school meals kitchen.

Due to the low load factor and hours of use, the energy requirements for heating and lighting should be modest in school buildings provided certain parameters are adhered to and users adopt an understanding attitude which will reduce, if not eliminate, waste when occupying the accommodation.

In order to maintain a reasonable level of environmental comfort to current standards, the building envelope, however effectively designed, needs a system for fine tuning the already "modified climate" within. The estimated energy needs for this system together with the requirements for catering and domestic services may be from 60 $kwh/y.m^2$ to 160 $kwh/y.m^2$ and to achieve a total consumption of energy nearer to the lower value of 60 $kwh/y.m^2$ it was considered essential to rely mainly upon natural ventilation and lighting, particularly during the late spring and summer,and to collect heat from ambient and renewable sources. Careful attention will be directed to the selection of windows and doors to ensure that air infiltration is reduced to a minimum and that natural ventilation can be finely controlled during the heating season whilst permitting liberal openings in summer time. Good air movement will be provided through roof mounted ventilators.

The investigation of ambient sources of energy included adjacent canal water, solar and wind. Although the use of the adjacent canal as a heat source for a heat pump system appeared at first to be viable, the scheme was discarded on grounds of poor cost effectiveness and doubts concerning system reliability due to the low rate of flow of canal at this point. Similarly the use of wind power for heat generation was not considered to be viable on what cost information was available nor was it possible to find a suitable site within a reasonable distance which would ensure a steady laminar air flow across the machine.

In all, eleven different schemes were considered on the basis of primary energy needed, nett present cost and system reliability. These are listed in figure 2. Undoubtedly the solution employing gas for supplementing the energy input is economic in terms of primary energy and operating costs but on the initial evidence available concerning reserves of finite energy resources the decision was to opt for off peak electrical supplementary heating. Thus (9), the solution chosen to provide the fine tuning to the internal environment, is anticipating energy trends in the next decade and would rely on partly solar and partly electrical energy input by off peak immersion heaters.

Consideration was given to the use of automatically controlled artificial lighting to supplement natural illumination and this is being installed and monitored in one experimental area for comparative purposes.

---

[1] also responsible for all monitoring on this project

SCHEME	Rankings		
	Nett Present Cost Capital and Running Costs	Prime Energy Requirement	Fuel and Operating Costs
1. Water/Water heat pump with electrical immersion heaters for domestic water heating	6	6	9
2. Water/Water heat pump with gas domestic water heater	8	2	5
3. Water/Water heat pump with off peak domestic water heating	5	6	8
4. Water/Water heat pump with off peak water heating and bulk storage	9	6	6
5. Conventional gas fire hot water space and domestic water heating	1	5	3
6. Air/Air heat pump with off peak domestic water heating	2	7	7
7. Air/Air heat pump with gas domestic water heating	3	4	4
8. Solar collectors, thermal storage and gas fired supplementary and standby heating	7	1	1
9. Solar collectors, thermal storage, off peak electric standby heating	4	3	2
10. Wind driven heat generator, thermal storage with gas supplementary and standby heating	11	*	*
11. Wind driven heat generator, thermal storage with off peak electric supplementary and standby heating	10	*	*

* Not included in ranking due to insufficient information available

Fig. 2 Schedule of energy input and utilisation systems investigated

ALTERNATIVE ENERGY SYSTEMS

Solar Heating and Hot Water System (Fig. 3)

The heat source will be a three part system using a solar panel, heat store and "OFF PEAK" immersion heaters. Heat storage will be contained in two water storage vessels one at low temperature and a second at a higher temperature. These will be designed to contain sufficient energy to satisfy the heating and hot water requirements of the school for a 24 hour period, with any deficit in the solar input being provided by "OFF PEAK" electric immersion heaters in the thermal store. Warm air heaters fed via a heat exchanger from the thermal high temperature storage tank will provide the heating requirements and will be

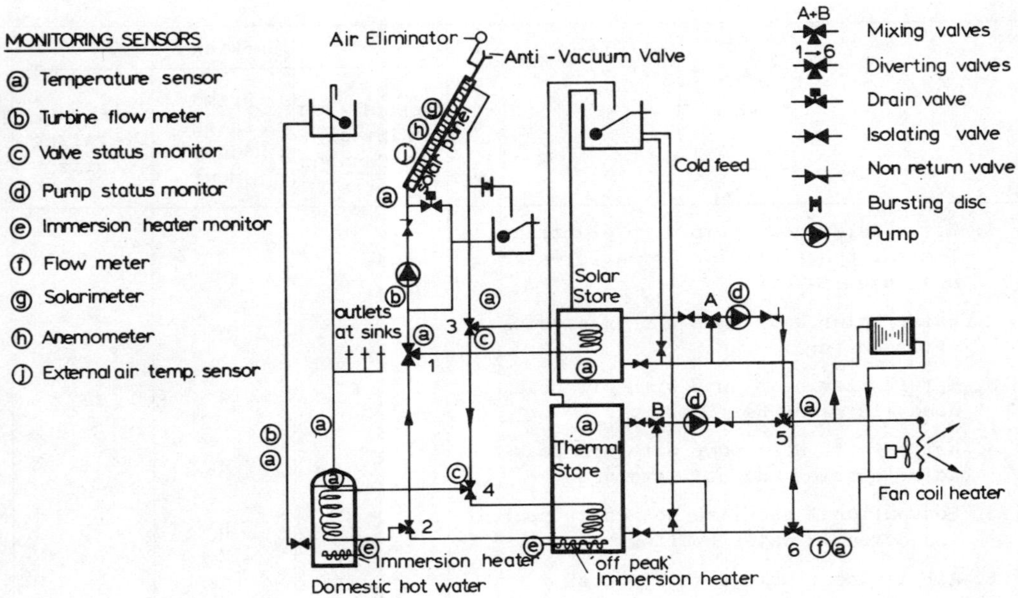

Fig. 3.

designed to raise internal temperature to 19°C in the occupied spaces.

Solar panels[2] of approximately 70 m^2 area will be installed on the roof at an angle of 33° and will be self draining in an approaching frost condition. Initially circulation is between the panels and the Solar (low temperature) store until such time as this store has been satisfied to a pre-determined temperature,[3] or when the temperature differential between the panel and the store temperature is small so that energy transfer is insignificant. The excess energy available is then offered to the thermal store by the operation of valves ref. 1 and 3, opening the ports to the thermal store and closing their ports to the solar store (valve ref. 2 and 4 also close the port to the solar store). When the thermal store has been satisfied then the excess energy is offered to the domestic hot water cylinder by the operation of valves 1 and 3 (closing their ports to the thermal store). In the event of both stores and the cylinder being satisfied the solar transfer pump will stop. The water in the panels will be transferred by displacement to the feed and expansion tank upon the operation of the motorised valve by the high or low temperature detector and the panels will drain.

The heating circuits are fed from the solar store and as the temperature in this store falls to a pre-determined temperature, valves ref. 5 and 6 will alter their open port position so that the radiator circuit is now fed from the thermal store. On a further fall in temperature of the solar store, valves ref. 7 and 8 will operate to alter the open port position so that the fan coil heater circuit is also fed from the thermal store. As the temperature of the solar store increases through the collecting period then the valves will reverse cycle to the preferred low energy using position.

---

[2] Type 36 black chrome collectors by Sunsense Ltd

[3] 95°C

Fig. 4. heat store

Fig. 5. fixing solar panels

Kitchen Waste Heat Recovery Units

These are devices, made up by the sub-contractor, for extracting waste heat, immediately beyond the traps, from the pipes containing discharge from sterilising and preparation sinks in the school Kitchen, and using it to preheat the cold feed to the H.W. cylinder. Figs. 6 and 7 show the unit and the method.

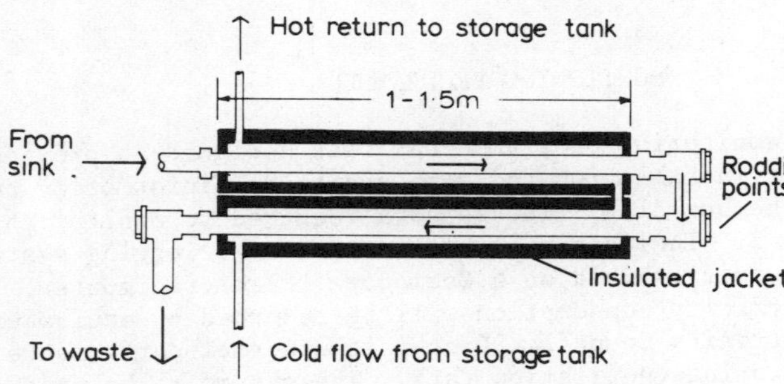

Fig. 6.

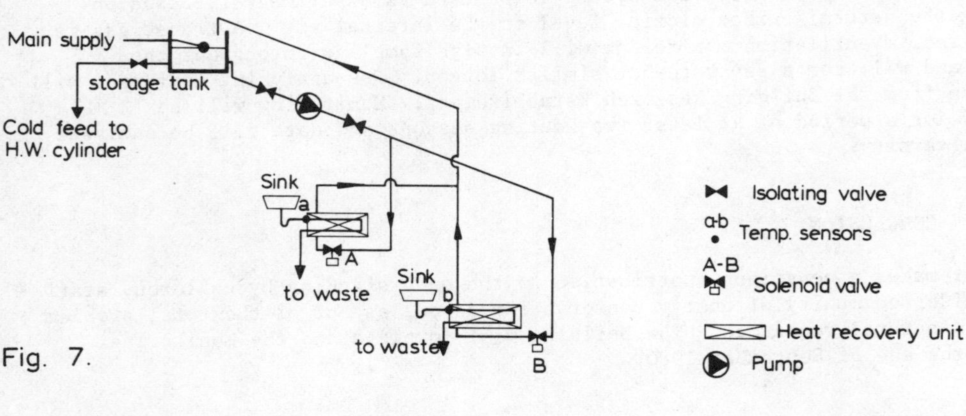

Fig. 7.

Rainwater Recycling Powered by Windpump

As an exercise in water conservation rainwater from the main roof areas is to be collected in an underground fibreglass tank. A windpump is used to raise this water to serve the W.C. cisterns in the school. Fig. 8 describes the method. It will be seen that excess water storage can be returned via the gutters to the collecting tank while inadequate rainwater supply can be augmented from the mains. A break tank ensures no contamination of the mains supply.

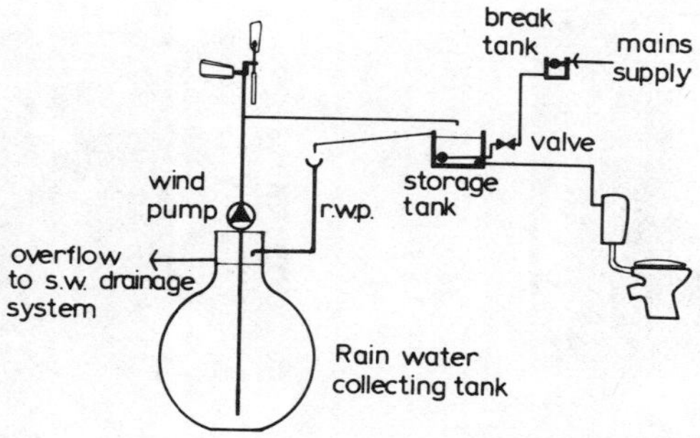

Fig. 8.

In order that the school may fully perform its educative function, services are exposed internally and colour coded so that the workings of the systems may be traced round the school. A mimic diagram located in the heat store and viewed from the main hall indicates and measures periods of solar input. Outside, the windpump is fixed to the west gable end and there will be opportunity for children to learn by direct experience about resource values of climate and soil fertility, and about cycles of energy conversion from sunlight to carbohydrates. Existing hedges will be retained, surrounding areas conserved as food and shelter for wildlife, and the children encouraged to relate to and respect the natural environment.

MONITORING PROGRAMME

The monitoring programme has been designed to provide detailed information on the solar installation and some additional information about the thermal performance of the building. The sensors required to monitor the solar system are shown on Fig. 3. The data will be recorded on a logging system developed at Exeter University, based on a Commodore PET microcomputer. Additional information about electricity consumption will be recorded by equipment on loan from the Electricity Council. Concerning thermal performance, emphasis has been laid on monitoring ventilation rate. The system to be used is experimental, based on recording the natural carbon dioxide level in the internal air. A tracer gas is not required. Ventilation monitoring will involve sampling throughout the building and will use a separate but similar logger. Gas analysing equipment will be on loan from the Building Research Establishment. Monitoring will be continued for a period of at least two heating seasons and data will be analysed at the University.

CONCLUSION

The school makes a conscious contribution to the understanding by children, staff and the wider community of energy conservation as one aspect of the total problem of limited natural resources. The building will complete and the monitoring start at the end of September 1980.

TOPIC F

Development and Planning

The Position of a Company Engaged in Small Scale Energy Development

T. H. Maguire

Sunderland House, The Dockyard, Pembroke Dock, UK

ABSTRACT.

The observations in this paper derive from the experience of the writer's small company which has been responsible for about 900 generator installations over the last 12 years. A few have been over 400 kW but the majority have been between 10 and 20 kW.

The area served is predominantly rural and bounded on 3 sides by the Irish Channel, and is 60 miles West of Swansea and 100 miles West of Cardiff.

The majority of users already have a single-phase supply from the South Wales Electricity Board which is generally reliable and at the time of installation adequate. Restoration of supply after a breakdown is rapid and efficient.

In serving the new market for small scale renewable energy equipment it is clear that a different category of user is involved with a new thinking about energy consumption patterns.

There is a great deal of spadework to be done in this field before it becomes profitable and it is unlikely that the user identified will be prepared to pay for a consultancy service.

KEYWORDS.

Diesel Generation; Tractor PTO Generators; Rural Industry; Peak Lopping; Consumption Patterns; Water Turbines; Leasing; Market; Windpower.

INTRODUCTION.

The object of this paper is to consider how a company already supplying small power plants can move into the field of renewable energy, and whether the experience gained with engine driven equipment is relevant. It has been necessary to determine the extent to which the same market is to be supplied with an alternative product, or whether an entirely new market is developing, and if the latter is the case, what additional problems can we foresee?

DIESEL.

Diesel generation is more costly than buying power from the Area Board except where sophisticated heat recovery arrangements are made, and we have to explain this to frequent enquirers who are seeking relief from their electricity bills. These enquirers are usually relatively small users of power, and small generators are less efficient than large ones. Another category of enquirer is building or rebuilding a house some distance from an existing power line and has been dismayed by the magnitude of the capital contribution sought by the Area Board to connect him. The capital sum required is frequently about the same as the cost of a generator installation, and we have to point out that a once-and-for-all contribution to the Board is a better proposition than owning a piece of machinery with all the maintenance and depreciation which goes with it.

Where a Board supply is not available for a domestic installation we normally offer a 'start-on-demand' unit such as Lister Startomatic. These however, require special arrangements if they are to provide the starting current required by equipment such as refrigerators. The limitations imposed by such sets are frequently unacceptable to people who have been accustomed to a mains supply, and are very similar to some of the problems encountered with small renewable energy units.

TRACTOR PTO GENERATORS.

About 500 of our installations have been on farms where electrical equipment has been installed, upon which relatively large-scale livestock enterprises are utterly dependant. These are commonly milking installations for larger numbers of cows than can be milked by hand in an emergency; not that modern cows will co-operate with hand-milking anyway. An interruption in electricity supply involving missing a single milking can have immediate financial repercussions and may slightly reduce the daily yield of the cows for the remainder of that lactation.

Farmers with tractors can usually manage with a power-take-off driven unit which is only about 50% of the cost of an engine driven set and is eligible for grant aid. We install a changeover switch which makes it impossible to backfeed into the Board's supply, and arrange with the Boards engineer to effect the final connection. Provided that the electricity meters are close together, we connect the farmhouse supply to the generator at the same time.

Controlled environment buildings housing intensive livestock enterprises are very sensitive to a break in power supply and the availability of a tractor-driven generator may not be adequate, particularly as its operation is dependant upon the presence of an operator. In this case, we offer a diesel set which will start up automatically upon failure of the main supply and take over the load in less than 10 seconds. Restoration to main supply can either be automatic or manual upon return of the operator. Fish farms come into this category and our experience as pumping engineers generally involves us in the whole installation.

Relatively few farms in West Wales enjoy a 3-phase power supply, and a growing proportion of farm equipment will not operate on single-phase, tower silo unloaders and reversible belt-feeders being the most common. The total amount of energy consumed by this 3-phase equipment is quite low and the Area Board may quote a very high figure for connecting a 3-phase supply, and moreover be unable to make such connection in time for the first season of use for the equipment.

In these circumstances, the farmer will buy a 3-phase generator and we will adapt his circuitry to provide single-phase standby power for his existing equipment. The farmer then becomes entirely dependant upon a single generator for 3-phase supply to essential equipment which must perform twice a day for 6 months of the year. He would be wise to have a second generator as a standby for the first, but in no case am I aware of this having been done. Instead he learns the writer's home telephone number and as we offer a 24 hour, 7 day service we cannot complain. In accordance with a well known law, nearly all breakdowns seem to occur between Friday afternoon and Monday morning.

The same factors which make it necessary for the farmer to be able to milk twice a day, 7 days a week, make it essential for the creameries which process the milk to offer the same continuity. In West Wales there are several large cheese factories, and a break in power supply not only prevents them from taking in the next days milk but interrupts their process, making it necessary to dispose of quantities of half-made cheese with the pollution problems that this can entail. We have supplied large 'mains-failure-automatic start' generators with the associated switch-gear in these cases, and where there is a computer-controlled operation we can supply a no-break system for the computer only.

RURAL INDUSTRY.

With the notable exception of the Milford Haven Refinery complex and the creameries, the rural industries of West Wales have often grown out of small engineering workshops in areas where there has only been a single-phase electricity supply and the installation of 3-phase equipment has demanded a generator for the same reasons as the farmer, but in a number of cases the user has progressively installed single-phase equipment until the Board supply is grossly overloaded and has then come to us for a solution. He will require a generator which will handle his existing equipment as well as his proposed expansion and we have built specially, single-phase generators up to 100 kW. and adapted his new, or new/secondhand plant to run on it. The unit cost of the power produced is high, but his overheads are already carried by his existing business and it is his only route to expansion.

In a number of cases, including some small saw-mills and quarries, the decision has been taken to install 3-phase equipment and the lead-time and the capital contribution required by the Area Board has been unnaceptable and they have installed generators as their main source of power. Where the average load is above 80% maximum, the cost of the electricity is comparable and if there is a requirement for hot water and this is recovered from the exhaust and cooling system of the generator, efficiency can be quite high.

PEAK LOPPING.

Electricity charges vary from area to area, but small users generally pay a flat rate which takes no account of peak demands, but the industrial user must pay a charge based on his ½-hour-maximum-demand, and it is here that there is some scope for the use of generators for peak lopping. The writer was involved in providing a standby generator for a small printing works after a break in power supply had caused a deadline to be missed. It was doubtful whether the cost of a standby generator to meet the maximum load was justified by the commercial risks involved but examination of the consumption pattern showed that 5 times as much electricity was used one day each week when they were melting type-metal and resulted in a very high maximum-demand charge. By using the generator on that one day, the maximum-demand charge is reduced by 80% and as the generator is running at full load it is quite economic. It is not a sound proposition to run the machine on the other 4 days and to do so might affect the supply contract with the Area Board.

CONSUMPTION PATTERNS.

Where a user has been accustomed to a mains supply of electricity, it is normal to find that average consumption is no more than 25% of peak consumption and may be very much lower, for instance, in the case of a hotel. The capital cost of a generator is related to the peak requirement but the machine is running for most of the time at a very low efficiencies. There are ways round this, but they cost money too.

There have been periods of political turmoil which have given impetus to the generator business, but it has been a very unsound base for developing a commercial enterprise, each settlement being followed by crop of cancellations leaving the supplier committed to accepting delivery of equipment. Almost as risky is the provision of equipment to cover small scale expansions as they occur. With each succeeding squeeze, whatever euphemism may be used to describe it at the time, the expansions are shelved for the time being. However, the greatest problem associated with this type of enterprise is that there is little repeat business and the better the supplier is at meeting the needs of the customer, the less likely is the requirement for maintenance or spares. People do change their cars every few years, and we are told that the average life of a house mortgage is 5 years, but a generator can last the business lifetime of the buyer. We have probably replaced less than 100 with larger units. Customer recommendation is quite another thing and is really the only source of new business.

There is quite a lot of D.I.Y. in the electrical wiring field and it is sometimes quite frightening to contemplate the system to which one must connect the generator. There is also a certain amount of D.I.Y. in generator installation and we write to those customers, advising them to adhere to I.E.E. recommendations. Amazingly, there is also a certain amount of D.I.Y. in generator construction. We do supply alternators, in particular, to such people with the same letter of advice. A variation on this is the customer who has the out of hours assistance of a 'man who works for the Board'. It is not always clear in what capacity he is so employed.

It should be apparent from the foregoing that the market for small generators is almost entirely to upgrade the supply from the Area Electricity Board, and usually at greater unit cost. This is not due to any shortcomings of the Board but to insure against supply failure, to provide 3-phase power where this is not available either in time or at an acceptable cost or to increase single-phase power beyond that which could reasonably be expected from the public supply. The only substantial exception to this is the provision of start-on-demand sets and it is significant that owners of such sets are often prepared to make substantial capital contributions to the Electricity Board to have a mains supply installed when it becomes available.

WATER TURBINES.

The Company is also involved in the supply of pumps and water turbines to industry, and therefore, has experience in this field. We have been appointed Agents for the MacKellar Water Turbine Generator throughout Wales and are negotiating with a number of prospective customers. The small water turbine has many of the characteristics of the 'start-on-demand' diesel set in that it can provide quite a lot of energy over a 24 hour period but is not suited to meet the high peak pattern of consumption sought by most of our existing customers. In fact, I can think of very few whose requirements could have been met by such an installation had the water been available.

We get a large number of enquirers for water turbine systems and the writer has been criticised for telling enquirers at an early stage that the cost is likely to be of the order of £1,000 per kW. installed. But it is the truth and results in about 90% of the enquirers losing interest. As existing categories of generator buyers will not be interested, perhaps we should consider the situation under which a person might install a water-turbine of less than 20 kW.

He must of course control a suitable site. Preferably an old mill where an earlier generation has done much of the spade work, perhaps literally, otherwise he must have a watercourse with sufficient water throughout at least 9 months of the year, and he should preferably control both banks. It is surprising how many prospective users only control one bank and have not considered the attitude of the other riparian owner. There must be a sufficient fall over a short enough distance to keep the cost of the leet and penstock acceptably low. A penstock may cost £30 - £50 per metre. It is essential that the leet and penstock should be sited on easily developed land as the cost of installation under rough rocky conditions is prohibitive. The site must be within 100 metres or so of the load, in order to minimise the length and cross-section of the cable necessary to avoid excessive voltage drop. Given that all this is available we come to commercial considerations.

ECONOMICS.

The standard of living which we take for granted in this country is sustained by an energy-intensive working environment. Let us consider whether the small water turbine is relevant to this and the extent to which it can contribute to a reduction of production costs. Let us further make the assumption that the whole of the output of the equipment can be applied to production for 40 hours in each of 50 weeks each year and that the remaining output substitutes 50% for electricity charged at full rate, and 50% at off-peak rates. Thus:- of 8760 generating hours in each year 2000 hours will relate to production time, 3380 hours will relate to a period when full rate will apply, 3380 hours will relate to a period when ½ rate will apply. Assuming that such a rigid energy consumption regime is acceptable the value of the output of the 3 available turbines is as follows:-

	Hours/ annum.	rate /kwH. £	200mm. Turbine. 3 kW. kwh	£	300mm. Turbine. 7 kW. kwh	£	500mm. Turbine. 19 kW. kwh	£
40 hrs. X 50 weeks.	2000.	.040	6000	240	14000	560	38000	1520
Full Rate.	3380.	.040	10140	406	23660	946	64220	2589
½ Rate.	3380.	.020	10140	203	2360	473	64220	1285
Totals.	8760.		26280	849	61320	1979	166440	5373

It is probable that electrical power which is available in such rigid and limited quantity has a lower value than that provided more flexibly by the Area Board, but I will not endeavour to assess the difference.

It is reasonable to assume that in a commercial situation, any cost representing less than 10% of total costs has a low priority in the allocation of capital and effort when economies are being sought. Thus, the possible commercial savings from installing a 3 kW. unit would only be relevant to enterprises with a turnover of £2400 to £8500 p.a. This would be a cottage industry which would be unlikely to be able to justify a capital investment of about £5,000 for this purpose. Such an enterprise might well not be registered under the V.A.T. regulations and would consequently be unable to recover the additional 15%.

The 7 kW. turbine will be relevant to enterprises in the £5,500 to £20,000 turnover bracket which will still be small craft industries such as potteries, health food mills and wood-working shops but it is still very doubtful whether a shrewd bank manager would consider a turbine the best use of about £6,000 of capital.

The 19 kW. unit is more attractive as a commercial investment, relating as it does to enterprises with a turnover in the £15,000 to £54,000 bracket, and costing about £8,000.

LEASING FINANCE.

It is probable that electrical energy will rise in cost in real terms by about 10% p.a. and that inflation will continue, perhaps at an average rate of 12.5% so it may be that an investment made at a time when there is only a break-even situation will become progressively more worthwhile as the years pass, however, there will probably be more urgent demands for capital from other sectors of the enterprise. With this in mind we are seeking to make arrangements, possibly through leasing, to finance the installations separately without competing for the limited capital available within the small business.

We hope that the cost in the first year will be no greater than the anticipated cost of fuels which would otherwise be purchased. The leasing costs in subsequent years would then remain fixed in cash terms while external costs of energy were rising by about 22.5% p.a. compound. At the end of the 5 or 7 years lease term, charges would be reduced to 1% p.a. of the original cost. It is unlikely that the smallest turbine could qualify for this arrangement except under unusually favourable conditions,

MARKET.

Those who have chosen to accept a modest life-style without outside subsidy are unlikely to be install turbine-generator equipment although they make up the majority of enquirers. Actual buyers are likely to have access to resources outside their immediate environment, perhaps by way of pension or other continuing interests but will probably have a genuine attraction to the principle of renewable energy which they can afford to satisfy. It will generally be a pre-requisite that the installation is economically viable but that it is quite a different thing from enjoying commercial priority over alternative uses for capital. It may even justify the user in his enjoyment of a very high level of domestic heating without making calls upon fossil fuels. In fact, the outstanding characteristic of small turbine systems is their need to apply the greater part of their output as low-grade heat. It may be that there are small scale commercial activities which can make use of this propensity which could be increased by the use of heat pumps if this were advantageous. We are offering complete units in kit form for the user to install as the cost of the civil work carried out at a distance would substantially increase the price, perhaps to a point where it became uneconomic. The potential user whom we meet is generally of a practical turn of mind anyway.

As energy costs rise, these small units are likely to become more popular and it may be that pressures will change from reducing unit costs by expanding production, to concentrating upon reduction of costs relating to a fixed output.

If we in the developed world are going to have to accept a steadily shrinking share of the world's output then the latter course may be consistent with it.

WINDPOWER.

The Company is evaluating the application in Europe of the Windstar high efficiency 10 kW. vertical axis wind generator developed in the United States.

It is planned to manufacture this unit in the communities where it will be used, predominantly on the West Coast.

Although a wind generator has some disadvantages, it also has the merit of being capable of replication once a suitable site has been found. There is likely to be some resistance to these installations on account of visual intrusion but the model which we are considering is aesthetically quite pleasing.

For remote areas without a mains supply we plan to offer the Windstar as a package in conjunction with a small diesel generator which will start automatically when the wind drops away, and as the wind turbine will generate A.C.current at 50 Hz it will be possible to synchronize both units to meet peak loads. However, the majority of installations will be on sites enjoying a mains supply and we will provide for automatic changeover from wind power to mains and back again, with all the necessary protection.

In England and Wales, synchronous co-generation with the Board supply is not at present acceptable and surplus energy will be diverted to a heat sink as with the water-turbine. High windspeeds also cause high heat loss creating a demand for the heat produced.

We must realise that the Area Electricity Boards have invested a great deal of public money in providing electricity supplies to remote areas and developments which make those supplies even less economic cannot be sound in the long run, particularly where the new system relies on the mains supply to back it up. I think that we should be aiming to supplement the main supply and improve the quality of life in remote areas with much higher standards of heating and other facilities than could be afforded at Board rates, but to plan to sustain mains consumption at present levels.

There are situations in the damp atmosphere of the West of Britain where a heating load can be provided throughout the year, and for this a wind generator is ideal. It is possible that we will build one of the first units on Iona after consultation with all concerned.

The high-grade electrical energy from the wind turbine can be multiplied through a heat pump system to provide 2 or 3 times the amount of low-grade heat and such an installation would be minimally affected by the lack of power during periods of low windspeed.

The erection of large groups of small wind generators to supplement the public supply involves quite different considerations and should be the subject of another study.

CONCLUSION.

The experience gained in providing a wide range of engine driven generating sets and designing installations to reconcile the users requirements with the power available, is likely to be invaluable in the renewable energy field, but it is apparent that the market is entirely different.

It may be that consumers of electricity will adapt their thinking to conform to the limitations of renewable energy sources but the writer does not expect this to occur. It is more likely that we must identify new areas which may absorb power from these sources and new people who are disposed to accept the disciplines imposed, in the cause of renewable energy.

There will, of course, be part loads which wind or water turbines can supply and in these cases we have come the full circle back to our current experience where a generating set supplements the Board supply.

There is going to be a great deal of investment of time and effort involved in moving into the renewable energy field and this will have to be carried by the existing business. Much of this effort will be related to site surveys in regard to which we shall endeavour to recover some of our out-of-pocket costs, but we do not anticipate that the customer will be prepared to pay an economic survey fee.

Alternative Energy and Regional Development

J. T. Hughes

Head of Policy and Research Division, Highlands and Islands Development Board, 27 Bridge Street, Inverness IV1 1QR, UK

ABSTRACT

This paper considers the links between alternative energy and regional development, specifically in the case of the Highlands and Islands of Scotland. My remarks on regional development and the role of regional development agencies are, however, relevant to a group of regions beyond the Highlands and Islands. The problem in preparing and presenting this paper is that both the topics of alternative energy and of regional development are extremely wide, with considerable overlap between them. I have, therefore, tried to focus on what I suggest may be the most useful points of collaboration and contact.

ROLE OF THE HIDB

After the numerous sessions in this conference on various types of energy issues and forms of generation, the participants will be much clearer about the field of alternative energy, both on the need for new solutions and the opportunities available to fulfil them. It might therefore be appropriate to, in a short space, outline the powers and functions of the Board as an example of a regional development agency.

The Board's powers can be grouped under four main headings.

(1) It can, at its discretion, give grants, loans and other financial assistance to people who want to set up or to expand businesses in the Highlands and Islands. It will also give advice on marketing, business management, production techniques and other support services.

(2) It can provide advance factories or workshops and build bespoke units for identified developers.

(3) It can undertake other projects and development schemes which it considers will benefit the social and economic life of the Highlands and Islands, and which are not likely to emerge from the private sector; for example, the Board has built hotels in remoter areas, mainly islands, to encourage the spread of tourism to these areas and to create a demand which will ultimately lead to development by the private sector. Another example under this heading would be the pilot agricultural marketing schemes to establish the

feasibility of improved marketing or production arrangements by individual farmers.

(4) Co-operation and liaison with the work of other relevant agencies operating in the Highlands and Islands. An example would be co-operating with local authority planning departments in the preparation of Structure and Local Plans; the planning powers are potentially important in establishing the framework for economic, social and community development both at regional and district level.

Many regional development agencies have primary responsibility for establishing the infrastructure in the form of social overhead capital and economic overhead capital which will allow development to take place. In the United Kingdom the provision of infrastructure is largely in the hands of local authorities and specialist functional boards. HIDB therefore has comparatively few powers to provide major items of infrastructure although it will, on occasions, co-operate with local authorities to provide finance for particular projects, especially in needy areas.

It is, of course, concerned that infrastructure provision is adequate in view of its important permissive role in stimulating development. Perhaps the best example of this concern is in the field of transport where the Board exercises advisory functions. When the Highland Transport Board was wound up in 1967 there was a recommendation in its final report that the HIDB be the principal adviser on transport matters in the Highlands and Islands to the Secretary of State for Scotland. The emergence of larger local authorities able to take a broader view of transport as compared to the old counties and with a statutory obligation to produce Transport Policies and Programmes (TPPs) has reduced this role. However, in a field like ferry transport which involves a wide number of government agencies and transport operators, the Board has undertaken research which it has disseminated publicly and most recently has produced ideas for a long-term restructuring of this area. [1]

If the Board is to become involved in areas where it has not a clear executive role, the model of transport policy would be how it should proceed, namely of establishing its concern with the area, the relevance to regional development, creating its technical capacity to understand the issues and, if it considers it necessary, to issue publicly ideas which it considers are not being given sufficient weight. Many people who have read our Transport Research Paper 6 do not agree with the suggestions therein; indeed it would be surprising if long-term changes were not to generate opposition - but surely any question is likely to benefit from discussion based upon forward thinking and specifically avoiding the assumption that the future will simply be a minor adaptation of the world as we know it today.

So far as energy is concerned the primary responsibility for provision lies with the main state boards for electricity and gas together with the role of the coal industry and major oil companies in providing most of the other hydrocarbon fuels. It is important to recognise the responsibilities of other levels of government and public agencies particularly when the concerns of a regional development agency spill over so readily into other functional areas. In consequence, much of the discussion between public bodies tends to be in private to avoid the distraction of a public dispute and to avoid the risk of fixing attitudes in defence of current policies. The desire not to tread on other toes is not simply a recognition of the territorial instinct of public servants and businessmen (although such motivations do exist) but it is clear that in many cases, to the outside observer, the divisions of responsibility between bodies do not square with their common sense perception of the problems. This is particularly true in rural areas where the economic and social conditions in an area are highly dependent on the interraction between job opportunities, housing, availability of services, transport and so on.

Since the field of regional development is wide, an important consideration for any agency like the HIDB is where to concentrate its effort and where to leave relevant work primarily to others. It is necessary to recognise the constraints on individual agencies set by the structure of the government. The HIDB would have powers to support experimental energy schemes mainly under the group of powers, outlined above. Those powers were exercised most clearly in the Colonsay case. However, the Board and its sponsoring department, Scottish Economic Planning Department, would wish to ensure that such a scheme was well founded technically, was particularly relevant to the Highlands and Islands, and that there were good reasons why other organisations specialising in energy economics or technology could not bear the whole burden. We have agreed with SDA to contribute financial support to a research project on sources of biomass material, conversion techniques and uses.

The whole question of energy costs of production, pricing and availability is a central question for an RDA. Some Regional Commissions in the United States are making planning for the future energy situation one of their central themes. It may be argued, for example, that a regional development agency should take an overall view of the efficient production and use of energy in its area but when there is reluctance by central government to establish an overall energy policy at national level as an important executive part of its activities, then it is clearly difficult at any sub-national level to deal strongly with that issue. There may also be advantages to a region in utilising some of the imperfections in the current system to its advantage and a need to point out that policies which may be justified on other grounds, eg the efficient use of energy may be damaging to particular areas. The proposal by Department of Transport to abolish Vehicle Excise Duty and to add that tax to the cost of petrol could have had an unduly damaging effect on rural areas where larger mileages are driven as an essential part of life and where the opportunities for substituting public transport are severely curtailed.

HIDB AND ENERGY

It might be useful to briefly outline some of the concerns with energy in the Board's work. You will have heard earlier from Alasdair Munro about the involvement of the Board with the Colonsay Electricity Association. At the other extreme from this local initiative, there will be discussion of the overall relationship between energy and the future of the area in the HIDB/Royal Society of Edinburgh-sponsored conference at Aviemore from 24/26 September 1980, immediately following this meeting. Looking to the future of major forms of generation, there has been contact with the Technical Advisory Group advising on the feasibility of wave energy which includes the Board's Deputy Chairman as member.

So far as small-scale energy is concerned, we have kept in touch with some of the thinking and discussion in this area but inevitably in a rather patchy way. Perhaps John Twidell has been the most prolific source of discussion and he was invited last year to give a seminar to Board members and staff. Various members of staff of the Board have held discussions and been involved in seminars on small scale energy. The largest single activity in this field was the commissioning of a research team under the auspices of the Church of Scotland's Society, Religion and Technology Project to look at 'Renewable Energy Sources in Highland Communities'. The results of this project have been subsumed under the papers on a number of other topics given at the seminar.

REGIONAL DEVELOPMENT AND ENERGY

If we are to think more comprehensively about the relationship between regional development and energy, we should be more specific about the links between the two subjects.

Energy can create employment which is one of the principal objectives of regional development. Ideally, the objective should be stated in terms of generating additional income since frequently small-scale development in remote areas does not create discrete full-time jobs but adds to the income generation of households, for example bed and breakfast tourist trade may add to the capacity of married women to earn money where there are few formal job opportunities for women.

(a) Most fundamentally energy development can make possible productive activities which were previously impossible through lack of any or an adequate supply of power. It is perhaps worth reminding ourselves of the debt the Highlands and Islands owes to the North of Scotland Hydro Electric Board in extending electrical power to a very high proportion of households in the area. Since its role, in common with other parts of the basic infrastructure, is largely permissive, it is difficult to claim that the availability of power 'created' employment opportunities. However, the 'availability effect' of such a universal requirement as electrical power cannot be underestimated. In view of its cheapness in many industrial heating and processing uses, the non-availability of grid gas and the higher cost of liquified supplies is a major disadvantage in some parts of our area.

(b) However, since energy is a cost of production, activities become more competitive if we have access to a cheaper source or one which is more compatible with specific production processes. The problem is that whereas the distribution of electricity and most forms of power is a centrally organised function, the needs of individual businesses and households are often highly specific, particularly in remote rural areas. The circumstances of individual households and some businesses may make them prefer a source of energy which is very cheap but potentially interruptable. Others are prepared to pay a premium for an assured, continuous supply. It may seem simple to allow each group to go their own way but they are interdependent. Since the cost of introducing and maintaining a centralised supply involves substantial overheads, the total consumption and its peaks have a direct effect on average unit costs.

A desire to stimulate development would suggest that there should be encouragement of cheaper forms of energy where these do not have major external effects on existing users. One of the difficulties in adopting a firm policy based upon this line is that the costs of conventional supplies are relatively predictable but those of new forms subject to wider variation and involve assumptions about the extent to which costs will fall as the R & D phase is passed.

(c) Energy generation can be a source of employment as a productive process in itself. The Board has supported the proposals to generate electricity by building a full-scale FBR at Dounreay. Another source of employment in direct production would arise from establishing wave energy capacity off the western coasts. In looking at some of the onshore implications of wave energy, as we have in a short report, it is clear that the generation process, involving the maintenance, supply and refitting of the units, would be a major industry in itself and that this would be a more direct local stimulus to employment than the possibility of establishing electricity-using industries in that area. The effect of the UKAEA in Thurso and its surrounding area has already created a most valuable stimulus to local infrastructure and other developments. There is, however, a limit in the extent to which a peripheral area can act as a producer of 'export' power where the capital cost of the grid and the transmission losses are high.

An RDA, therefore, must press for the availability of power supplies at competitive prices within its area. Alternative energy production would make three types of contribution. (i) Major new forms of generation could increase employment and attract energy-using industries. Any northerly shift in the centre of gravity of production must make the area less peripheral. Realistically it will be the end of the century before such a shift is possible. (ii) At an intermediate stage, new methods of production can bolster specific weaknesses of the current system, specifically the opportunity to identify lower cost alternative supplies to lower the cost of generation in areas dependent on high cost methods such as islands with diesel generation. At the very least the development of alternative systems may present options for future investment. The decision of the Hydro Board to install a windmill in Orkney in apparent pursuit of such a policy is to be welcomed. Other options are tidal schemes especially where the cost of the barrage could help provide a road short-cut. The lack of tidal range will be a major disadvantage unless turbines can operate efficiently with lower pressure differences than previously considered.

(iii) Thirdly, small-scale alternative systems for individual households/workshops (or in small groups) could lower costs. Possibly the most promising ideas are small windmills or hydro units. Both have problems of continuity and of controlling the cycle phasing. Just as the Board has helped provide diesel generators as part of the capital required for certain businesses, so it would be prepared to consider other schemes if they can be shown to reduce the costs of operation. I understand one of the important issues in the development of small-scale schemes within the grid area is the interchange of power with the grid system. It will be obvious why the NSHEB or any other electricity authority will have reservations about free interchange - greater uncertainty about consumption of their own power, additional switchgear, less control of technical specification. If there is greater diffusion of small units, the possibility of establishing the manufacture of such equipment in our area increases. The Board has already assisted the McKellar Engineering business in Grantown. Generally, however, one would expect a great deal of such manufacturing would be linked to general engineering businesses.

ENERGY IN A PROGRAMME OF REGIONAL DEVELOPMENT

Current policy suggests that it is unlikely in the UK that low energy costs in one region would be used as a strong incentive in the attraction of industry to that region from the rest of the country or internationally. The major incentives will generally be the grants, loans, factory building and advice services mentioned above. It is essential, however, that no one region is singled out by higher energy costs in a broadly uniform pattern. At present large industrial users in island areas pay a surcharge on consumption over 500,000 units annually, a threshold which was recently lowered from one million units. The HIDB would obviously wish to see that disadvantage eliminated and competitive prices offered for major industrial users in the main grid area, but have to recognise the commercial implications for the NSHEB. Part of the difficulty in fitting any assumptions about the price of large-scale electricity supply into a promotional programme is the lack of knowledge about the underlying costs of production which are commercially confidential to NSHEB.

How can the development of a region benefit from the availability of indigenous energy resources? Many of these resources are developed by organisations with a national or international orientation whose objectives take little or no note of regional interests. I would hope that when new forms of electricity generation are developed in the Highlands and Islands, the supplies, eg from the FBR and Wave Energy, will be made available during the development phase at agreed prices related to the cost of 'conventional' sources of generation. Although in the UK there have been few examples of energy supplies and pricing being used as a stimulus to development, there is a much stronger case for such a policy in the Highlands. Here the major contribution of the aluminium industry would not exist without firstly the ability of the private sector to secure cheap hydro supplies and subsequently the government-subsidised (obviously beyond contemporary expectations) prices guaranteed in the 1960s. In cases where location in the Highlands and Islands has other advantages for an industry, the Board would argue for a favourable energy price which reflected the social and economic benefits.

Some indigenous resources can be used for energy or other purposes. An objective of HIDB has been to stimulate the development of peat for energy use, eg by fluidised bed combustion, for chemicals, eg coking peat, or for horticultural use. The prospects depend in great measure on the relative prices of the final products in which the resources are used. The Board and the Scottish Development Agency are co-operating in an EEC programme to examine the potential of biomass methods in energy generation. One of the major planks of the Board's development strategy in recent years is the use of marginal North Sea petroleum resources as chemical feedstock rather than in lower value energy uses from flaring to burning in electricity generation. Specifically we have promoted the use of natural gas liquids in petrochemical processing at Cromarty Firth.

It is tempting to concentrate on new developments but the prosperity of a region over the next generation will largely depend on the competitiveness of the existing structure of economic activity. There should be a major programme to improve energy utilisation in all sectors from agriculture to services. Land Development Division and my own in the Board are currently considering a project to examine the opportunities to improve energy efficiency in that sector. Part of the reasoning behind the HIDB fisheries strategy was that improved landing facilities in the Western Isles, eg Breasclete, would reduce the energy requirements of fishing especially in exploiting new pelagic species in the Atlantic. Tourism tends to be an energy-intensive activity especially in the case of car touring holidays. Since it continues to show a high income elasticity, it is a long-term growth sector in which it should be possible to encourage greater use of packages with public transport and single centre holidays. Of 'alternative' solutions a positive conservation policy could well make the greatest contribution to regional employment. Gerald Leach and Associates have examined ways in which energy demand could be contained, it remains to be seen in detail how far the reduction in energy inputs has to be compensated for by substituting other factors of production.

The greatest and least tractable influence on energy consumption in a region is the settlement pattern. It directly determines the provision and operation of infrastructure, the transport costs and domestic energy consumption. The traditional pattern has one bonus by creating access to domestic peat cutting. We have assisted in an experimental scheme to adapt peat-cutting tractor attachments but the task is proving difficult. In general, higher energy prices raise the cost of movement to and within a sparsely populated peripheral area. It would be premature, however, to draw the conclusion that we should be any less determined to protect these areas. It may perhaps be an opportunity to rethink the reasons for and form of our policies. It is not clear to me that people in wide areas of the Highlands receive transport and other subsidies at a higher rate than inner city or even prosperous suburban areas; the suburban motorist or railway commuter escapes many of the real costs of his actions.

It is possible to think of a number of ways to improve the welfare of people in the Highlands and Islands. Support for a network of services would pay off by reducing personal transport costs. It has been suggested that existing mobile services and some additional ones might be co-ordinated to create mini-market days by being in the same communities at the same time.

The ultimate object of economic development, the Board's primary concern, is to improve the welfare of people in the area and increase their contribution to national prosperity. The development of alternative energy can contribute to that object but priority would obviously be assigned to projects with a fairly definite promise of achievement as compared to more conventional policies.

REFERENCES

1 See Transport Research Paper 6 - The Future of Ferry Services in the Highlands and Islands - available from HIDB, postage 22p.

Government Policy and Alternative Technologies in Rural Areas Participation Group Discussion

Reporter: **Alex Walker**

Glasgow University, UK

Participants:	David Crabbe	Open University
	John Duncan	Orkney Island Council
	Alastair Fraser	International Forest Science Consult.
	May Johnston	H I D B, Inverness
	Owen Lewis	University College, Dublin
	Alan Mowle	Nature Conservancy Council
	Alastair Munro	H I D B, Lochgilphead
	Joan Richardson	N S H E B Consultative Council
	Malcolm Slesser	Strathclyde University
	Kevin Woodbridge	North Ronaldsay

The driving force behind this group was the feeling expressed by many speakers at the Conference that government at all levels is failing to provide adequately for the development of alternative technologies in rural areas. The mood of the participants was generally a positive one with many suggestions for aid and action being brought forward and a high degree of consensus being reached on almost every issue. The proposals fall into three broad categories:

a) Administrative
b) Financial
c) Public Relations

ADMINISTRATION

It was noted that one of the greatest difficulties to the development of new energy technologies is the vertical organisation of government departments and agencies. What may be needed is a co-ordinated approach, preferably at local level, to examine the potential for energy production. The competitive and bureaucratic natures of the various power boards, and the Department of Energy's concentration on global solutions to national problems are strongly militating against cost effective and efficient energy programmes. A single energy utility on a regional basis could be the optimum solution, possibly under the aegis of the Regional/Island Councils. Ideally the Forestry Commission, transport organisation and other energy suppliers/users not normally considered in the energy equation should be included, and information regarding the costs and performance of new technologies should be gathered and collated.

Further to this, on a more specific note, the Electricity Boards come in for some criticism. Firstly the remit of the 1947 Government Bill which set up the NSHEB, contained provision for the encouragement of energy supplies of all kinds and the lack of action on this front has been a major hindrance. Secondly the difficulties encountered by energy producers in selling electricity to the national grid was seen as a major disadvantage.

FINANCIAL CONSIDERATIONS

The miniscule level of government funding for alternative technology is a well known fact but needs continual emphasis. Many jealous eyes were cast at the "Ark" project which received $6 million from Canadian government sources, and Dr Todd's comment that the UK government spends more on military brass bands than on renewable energy sources was in the forefront of everyone's mind. Whilst the thrust of further financial support must come from central funds it was generally considered that the HIDB could spend more of its (admittedly limited) resources on the subject and that the local authorities, particularly the Shetlands Island Council, might be persuaded to take an interest.

A number of other points were made - that present government policy prices energy resources with regard to the cost of extraction, not the cost of replacement, and even a limited change in the latter direction would aid renewable sources,

- that the high initial cost of installation of new technologies should be borne in mind by government and that grants for this should be more readily available as maintenance/running costs of experimental projects can be more easily borne by small concerns,
- that the cost per job created by conventional technologies such as nuclear power and petrochemicals is often vast in comparison to alternative possibilities.

PUBLIC RELATIONS

Sympathy was expressed for the opinion that a change of attitude from both the lay community and government is just as important as and indeed a prerequisite for further financial aid. It was noted with regret that there were no representatives from the Western Isle Council or Argyll District, and that the Highland Regional Council had shown minimal interest. Pressure must be brought to bear on these bodies to make them aware of the urgent need for their co-operation in ventures involving new energy sources. The increase in oil prices and the recent attempts by the NSHEB to impose a surcharge on remote communities should make the arguments in favour of their participation persuasive. It is not enough to be enthusiastic amongst ourselves, we must transmit our enthusiasm to others if we are to achieve what we week.

CONCLUSIONS

There is an urgent need for central government to investigate all aspects of energy provision in rural areas, in particular to co-ordinate the presently desperate actions of the various supplying bodies and to help with the testing of new technologies. Secondly there are a number of ways in which government could act to lessen the problems facing experimental projects, in particular the removal of obstructions to the sale of electricity to the national grid. Thirdly, the level of government support for new energies technology is derisory and must be increased if the Highlands and Islands and other rural areas are to play the role in energy generation they are capable of doing. Finally, it is not enough for alternative technologies to develop better methods of resource utilisation, there must be a concentrated and continuous attempt to influence government policy in a desirable direction.

Practical Application of Wind Power to Rural and Island Communities
Record of Participation Group Discussion

Reporter: **David Wheeler**
Participants: **S. Bannister, S. Gair, P. R. Myers, A. Stobart and J. Towrie**

The title of this Group was suggested in order to stimulate discussion of what could or ought to be done to make wind power applications to rural communities a practical proposition. It was felt that, at the present time, there was no general movement to this end, only the efforts of a few individuals or small groups working more or less alone and usually on a small scale. Mr John Towrie Orkney Islands Council, pointed out that this body was looking into the possibility of heating the North Ronaldsay school by means of a wind driven heat pump. Mr Wheeler reported that the Fair Isle Electricity Committee was seriously considering a wind turbine generator for Fair Isle. Though there was a need for assistance in these and other projects, there was as yet no one unbiased body to which to turn for help. Choice of equipment was also difficult, with no independent advisory body to put forward recommendations. Dr Twidell's suggestion of a central test centre was thought to be a very good idea, but its location would need to be chosen with care. The area must be one with a high wind regime, but yet be easy of access - Orkney was one of several possibilities. Another idea might be to have a central monitoring facility linked to wind turbine generators positioned in less accessible places.

Wind has been used to provide mechanical power for centuries, but its application to the generation of electricity is only a relatively recent development. As such, there is still a credibility gap to bridge before responsible bodies will provide financial support directed towards this application. Island and remote rural communities are at present importers of such developments - if the peripheral areas required less fuel, more would be available for the central areas. It was the group's opinion that, rather than subsidising rural and island communities, more should be offered in the form of grants. With such grants, it should then be possible for many islands and communities of the West and North coasts of Britain to become independent of the rest of the country for their energy needs. To achieve this, these islands and communities will need to make known to Central Government their needs and suggested solutions - probably the greatest difficulty of all.

List of Participants

AGNEW, Mr Patrick W.	Department of Mechanical Engineering University of Glasgow. GLASGOW.
AUER, Miss Christa	Heliotechnic Associates International 5 Dryden Street LONDON WC2E 9NW
ADAMS, J.D.	10 Gordon Place, Withington MANCHESTER.
BAKER, Mr Michael	Energy Research Group Open University Milton Keynes, MK7 6AA
BALLARD, Mr L.J.	ERA Technology Ltd Cleeve Road Leatherhead SURREY KY22 7SA
BANNISTER, Dr Scott	Department of Mechanical Engineering Napier College EDINBURGH.
BARBOUR, Mr David	Department of Applied Phusics University of Strathclyde GLASGOW
BOSLEY, Mr K.	American Refinery Company WTG Energy Systems Inc York Road Sparks MARYLAND 21152, U.S.A.
BURGESS, Mr John M.	Director of Research & Development 93 St Olaf Street LERWICK, Shetlands
BURTON, Mr Anthony	1 Attadale Road INVERNESS LV3 5QH
BURTON, Miss Wendy	Fiat Totem Ltd Great West Road BRENTFORD TW8 9DJ
CAMPBELL, C.J.	Highland Heating Savers, Midfearn, Ardgay, Ross-shire.
CAMERON, Mr Richard	Deputy Director Planning Department Highland Regional Council Glenurquhart Road INVERNESS

CARRIE, Rev John	South Queensferry Church South Queensferry EDINBURGH
CLARKE, Mr James	The Shian, Lochloy Road NAIRN
COCKBURN, Mr Harry	Stanley P Morrison Ltd 13 Royal Crescent GLASGOW G3 7SU
CRABBE, Mr David	Energy Research Open University Milton Keynes, MK7 6AA
CRICHTON, Mr Charles	Monzie Castle Crieff, PERTHSHIRE PH7 4HD
DOBSON, Mr Terry	Project Development Officer Scottish Development Agency 120 Bothwell Street GLASGOW G2 7JP
DOUGLAS, Mr G.A.Brown	Mansefield Bents Road, Chapelton by Strathaven, Lanarkshire.
DUNCAN Mr John	Department of Planning and Development Orkney Island Council KIRKWALL, Orkney.
EISNER, Professor E.	Applied Physics Department University of Strathclyde John Anderson Building GLASGOW.
FRASER, Dr Alastair I.	International Forest Science Consultancy 21 Biggar Road Silverburn PENICUIK EH26 1LQ
GAJR, Mr S.	Department of Mechanical Engineering Napier College Colinton Road EDINBURGH
GILCHRIST, Mr Timothy	Architect, Mavisbank Newrath WATERFORD, Ireland

HAMILTON, Mr A.	Shetland Islands Council Victoria Buildings, The Esplanade LERWICK.
HUGHES, Mr J.T.	Gead of Policy & Research Division Highlands and Islands Development Board Bridge House 27 Bank Street INVERNESS IV1 1QR
JACQUES, Mr Robin	10 Gordon Place Withington MANCHESTER M20 9LR
JOHNSTONE, Dr May	Highlands and Islands Development Board 27 Bank Street INVERNESS LV1 1QR
KNIGHT, Mr Anthony	Ardullie Lodge DINGWALL, Ross & Cromarty.
KUMAR, Mr Mahendra	School of Agriculture Sutton Bonington Loughborough Leics LE12 5RD
LARKIN, Mr Stephen	National College of Agriculture Engineering Silsoe BEDFORD MK45 4DT
LAWSON, Mr Gerald	Institute of Terrestrial Ecology Merlewood Grange-over-sands CUMBRIA
LEWIS, Mr Owne	School of Architecture, Earlsfort Terrace DUBLIN, Ireland
LIPMAN, Professor N.H.	Rutherford Laboratory Chilton DIDCOT Oxfordshire
LOW, Mr James	General Technology Systems Ltd Forge House 2o Market Place, BRENTFORD Middx.
LOWTHER, Mr Bryan	North of Scotland Hydro Electric Board 8 Albert Street KIRKWALL, Orkney KW15 1HP

McCHLEARY, Mr Alan	APD Field Manager BP Oil Ltd 3 Ardross Terrace INVERNESS IV3 5NH
MACDONALD, Mr H.B.	North of Scotland Hydro Electric Board 16 Rothesay Terrace EDINBURGH EH3 7SE
MacGREGOR, Mr A.W.K.	Department of Mechanical Engineering Napier College Colinton Road EDINBURGH EH10 5DT
McKAY, Dr K.T.	Director, The Ark Project Rm 4m Souris P.E.I., CANADA COA 2BO
McKEOWN, Mr John	McLeod & Miller Engineers Ltd Whistleberry Road Blantyre GLASGOW G72 OTG
MacELHOSE, Mr D.	National Trust for Scotland 5 Charlotte Square EDINBURGH
MAGUIRE, Mr T H.	Managing Director, Tom Maguire & Co Ltd Sunderland House The Dockyard PEMBROKE DOCK.
MALCOLM, Mr S.	Sir M MacDonald & Partners 77 Church Street INVERNESS IVL 1ES
MARKS, Mr Brian	17 Banchorn Avenue Inchinnan, Renfrewshire PA4 9PX
MASON, Dr David	P.V.A.F. Ltd 7-8 Kendrick Mews LONDON SW7 3HG
MAYS, Dr Ian	P.I. Specialist Engineers Ltd The Dean, Arlesford HANTS

MIDDLETON, Mr Norman	Director, Regional Insulation Services Ltd 42 Viewforth Terrace EDINBURGH EH10 4LJ
MITCHELL, Dr Paul	Forestry Department University of Aberdeen ABERDEEN
MORRISON, Mr Michael	University of Cambridge Department of Phusics Cavendish Laboratory Madingley Road CAMBRIDGE CB3 OHE
MOWLE, Mr A.	Nature Conservancy Council 12 Hope Terrace EDINBURGH EH9 2AJ
MUNRO, Mr A.	Highlands and Islands Development Board Kilmory LOCHGILPHEAD Argyll
MYERS, Mr Philip R.	Highlands and Islands Development Board 2 Bank Lane LERWICK, Shetland
OTHIENO, Mr Herick	Department of Applied Physics University of Strathclyde John Anderson Building GLASGOW
POPE, Mr Ian	North of Scotland Hydro Electric Board 16 Rothesay Place EDINBURGH EH3 7SE
PRINGLE, Mr Robert	North of Scotland College of Agriculture ABERDEEN
REID, Mr S.	2 Wester Raddery Fortrose Black Isle Ross-shire IV9 8RJ
REILLY, Mr	Perlit Engineering Development Ltd Balgay House Inchture PERTH.

RIDDOCH, Miss Fiona	Department of Physics Heriot-Watt University EDINBURGH EH17 4AS
ROSS, Mr Alistair	Production Director Stanley P Morrison Ltd 13 Royal Crescent GLASGOW G3 7SU
SCHELLENBERG, Mr Keith	Maybank Udny ELLON Aberdeenshire AB4 OPQ
SCHNADT, Mr L.	Director New Findhorn Directions Ltd The Park FORRES, IV36 OTZ
SEXON, Miss Barbara	Department of Engineering University of Reading Whiteknights READING, Berks.
SELIGMAN, Mr R.	National Trust for Scotland 109 Church Street INVERNESS
SHARPE Mr	Managing Director Perlit Engineering Development Ltd Balgay House Inchture PERTH
SHAW, Dr T.L.	Sir Robert MacAlpine & Co Ltd 40 Bernard Street LONDON WC1N 1LG
SLESSER, Dr Malcolm	Director, Energy Studies Unit University of Strathclyde 100 Montrose Street GLASGOW
SMALE, Mr T.	Fiat Totem Ltd Great West Road BRENTFORD TW8 9DJ
STOBART, Mr Andrew	Director Trimble Windmills 55 East Parade Harrogate N.Yorkshire HG1 5LQ

TAYLOR, Mr Julian	Atkins Research & Development Ashley Road EPSOM, Surrey
THOMASON, Mr Edward	Shetlands Islands Council Planning Department, Victoria Buildings The Esplanade LERWICK
TODD Dr R.	Centre for Alternative Technology Machynlleth Powys Wales
TOWRIE, Mr John	Orkney Island Council 11 Broad Street KIRKWALL, Orkney
TWIDELL, Dr John W.	Applied Physics Department University of Stathclyde John Anderston Building GLASGOW
WALKER, Mr A.	Department of Management Studies University of Glasgow 25 Bute Gardens GLASGOW.
WARNE Mr	ERA Cleeve Road Leatherhead, SURREY
WATSON, Mr	Northumbrian Energy Workshop Hexham
WHEELER, Mr David	Field Croft Fair Isle Shetland
WHITE, Mr Roger J.	Alternative Technology Centre Machynlleth, Powys Wales
WHYMAN, Mr Peter	Department of Architecture University of Strathclyde GLASGOW

WOODBRIDGE, Dr Kevin F.	New Manse North Ronaldsay ORKNEY, KW17 2BE
WYPER, Miss Hilary	2 Cameron Crescent Carmunnock Glasgow.

Author-Paper Index

(See participants' list for addresses and institutions)

continued:

* The address at the Conference Dinner was made possible by financial support from Occidental of Britain Inc. from their Kirkwall, Orkney, office.

Subject Index